No. 3175
$23.95

Understanding Lasers

Stan Gibilisco

To my parents
Who helped make this book possible

FIRST EDITION
FIRST PRINTING

Library of Congress Cataloging-in-Publication Data

Gibilisco, Stan.
Understanding lasers / by Stan Gibilisco.
p. cm.
Bibliography: p.
Includes index.
ISBN 0-8306-9275-4 ISBN 0-8306-3175-5 (pbk.)
1. Lasers. I. Title.
TA1675.G53 1989
621.36'6—dc20 89-31852

CIP

TAB BOOKS Inc. offers software for sale. For information and a catalog, please contact TAB Software Department, Blue Ridge Summit, PA 17294-0850.

Questions regarding the content of this book should be addressed to:

Reader Inquiry Branch
TAB BOOKS Inc.
Blue Ridge Summit, PA 17294-0214

Roland Phelps: Acquisitions Editor
Daniel Early: Technical Editor

Paperbound cover photograph courtesy of Coherent, Inc., Palo Alto, California.

Contents

Introduction

Lasers have received a great deal of publicity since they were first discovered. Their ability to propagate across long distances with almost no attenuation, and their intensity, which allows them to be used as "ray guns," created a sensation. Wild visions developed almost overnight: powerful weapons that would forever change the art of making war, that would enable us to communicate with alien civilizations, that could vaporize lakes and even whole oceans, that could blast killer meteorites before they could annihilate all life on Earth—these and other scenarios were exploited by the science fiction community. Promises were made that cancer might be cured, diseases conquered, production costs slashed. Some of these dreams are being realized, but the practical scientist knows that some of them must remain fiction, at least for a few more generations.

Hardly a year goes by in which some major discovery is not made that involves these devices. Lasers have been used for purposes as diverse as treating cancer, blinding reconnaisance cameras, and conveying pictures via optical fibers. Laser weapons are in the developmental stage, and the coherent light from lasers is proving itself useful in ways we have only begun to imagine. Lasers can be found everywhere today, from the grocery checkout register to elevator doors and burglar alarms. Laser diodes can be bought for pennies in electronics hobby stores. Lasers are even used in the entertainment industry, in such things as light shows, discotheques, and holograms.

In this book, we will examine the nature of laser light and the ways it can be generated. Then we will explore some of the many uses of this device. We will examine how lasers can be used in place of scalpels in surgical operations to minimize trauma, bleeding, and the risk of infection. We will see how lasers can be modulated to convey large volumes of information at high rates of speed, both via optical fibers and directly. We will see how holograms can be made, and how lasers can be used to defend against nuclear attack.

Laser technology is rapidly advancing, and there will be things that this book does not mention. All technology is moving ahead at such a pace that we sometimes feel we cannot keep up. Nonetheless, if you have an interest in laser applications, you might get some good information and ideas here, so you can pursue your interest in more detail. Specialized texts in various laser applications are recommended for the serious laser enthusiast and scientist.

If this book whets your appetite for more, so much the better.

1

The Concept of the Laser

Most of us probably think of lasers as powerful ray guns, capable of boring holes in solid steel or concrete, or shooting down planes and satellites. Lasers are powerful devices, but the notion that they are all like knives of light is not really very accurate. Lasers have a wide variety of uses and there are distinctly different kinds of lasers. In fact, very few lasers are powerful enough to weld, bore holes, or act as ethereal "star wars" missiles.

Laser is an acronym for the long term, Light Amplification by Stimulated Emission of Radiation. Originally, the word *light* was *microwave* and the device was called a *maser*. The laser, developed somewhat later, was originally known as the *optical maser*.

I recall playing (the high-school version of experimenting) with the helium-neon laser in the physics department long after school on winter afternoons. The sun set at about 4:15 P.M. in Minnesota in December, and by 5:00 it was nighttime. I would shine the laser out the window onto reflective street signs blocks away and marvel at how bright the signs became. I would set the laser up in the long hallways and measure the diameter of the bright red spot at the other end of the corridor. I tried using color filters and wondered how the blue piece of cellophane could filter out the laser light so totally that I could look directly into the laser itself and see how it was put together. (*Warning*: This is not a safe experiment. Don't be as foolhardy as I was!) I wanted to figure out how I could modulate the beam so I could transmit signals over long distances, possibly using reflecting mirrors at various points to create a closed-circuit system with a friend across town who was not in direct line of sight with my bedroom window. The janitors and teachers who stayed after school knew me well; I had done other experiments in chemistry and this behavior was no surprise to them. I wonder if any of them thought I might turn the laser on them and burn a hole in their shirts or pants?

ELECTROMAGNETIC WAVES

Light energy, and in fact all forms of radiant energy, take the form of waves. We can think of electromagnetic energy—radio waves, infrared, visible light, ultraviolet, X rays and gamma rays—as a barrage of particles called *photons*, or as a wave disturbance. This particle/wave dichotomy is a part of physics that disturbs some students of the subject, because radiant energy displays properties of both the particle and the wave. The energy contained in a given particle is in direct proportion to the frequency of the wave disturbance. The higher the frequency, the more energy is contained in a given particle. Exactly what these particles look like and how big they are is not known, just as we don't know what an electron, proton, or neutron would look like if we could enlarge it enought to put in on the dining room table as a centerpiece. To understand the laser, it is best to use the wave model of electromagnetic energy.

An electromagnetic disturbance may take the form of a sine wave (FIG. 1-1), in which case it has a definite, indentifiable wavelength λ and a frequency f. The wavelength in meters and the frequency in Hertz (cycles per second) are related by the formula:

$$\lambda = 3 \times \frac{10^8}{f}$$

For example, if the frequency is 10 MHz or 10^7 Hz, the wavelength in meters is $3 \times 10^8/10^7 = 3 \times 10^1 = 30$. This is the formula for wavelength versus frequency in a vacuum, or "free space." It is essentially the same in air. In solid, transparent materials such as glass, plastic or water, the relationship is different, the constant often being considerably less than 3×10^8. The converse formula, giving frequency in terms of wavelength, is

$$f = 3 \times \frac{10^8}{\lambda}$$

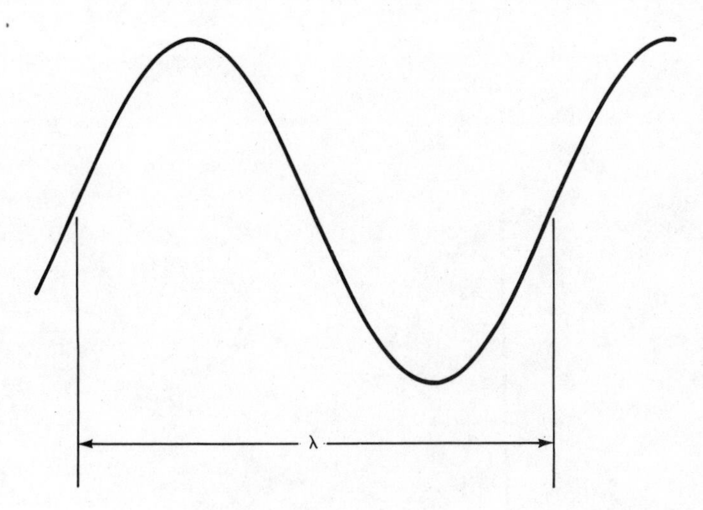

Fig. 1-1. A pure sine wave.

In the case of a broadcast station, the carrier wave without modulation is theoretically a single-frequency, single-wavelength signal. There are invariably some harmonics, or multiples of the fundamental frequency, but in theory the signal has just one frequency. In practice there is a certain amount of noise on the signal, spreading the frequency range of the signal so it has a bandwidth of a few Hertz instead of zero. But in the ideal case we might show the spectrum of the signal as a single line on a graph depicting the relative amplitude, such as output power in watts. An example is shown in FIG. 1-2. The unmodulated carrier would look like the sine wave of FIG. 1-1 if displayed on an oscilloscope.

Once the carrier is modulated, the bandwidth is no longer zero, even in pure theory. Suppose we modulate the signal with "white noise" having a characteristic amplitude-versus-frequency curve as shown in FIG. 1-3A. The amplitude-modulated signal would have a spectral distribution like that in FIG. 1-3B. The amplitude of the "sidebands," or components of the signal on either side of the main carrier, would depend on the percentage of modulation. But the signal would occupy a certain amount of spectrum space and the wave would no longer have the clean, sinusoidal shape of FIG. 1-1. If the signal were displayed on an oscilloscope, it would appear as a complicated jumble of waves, constantly changing phase.

Visible light, as we see it, is not normally a clean, sine wave kind of electromagnetic disturbance like an unmodulated radio wave. There are components from the red, or longest visible wavelength, through the violet, or shortest visible wavelength. Red light

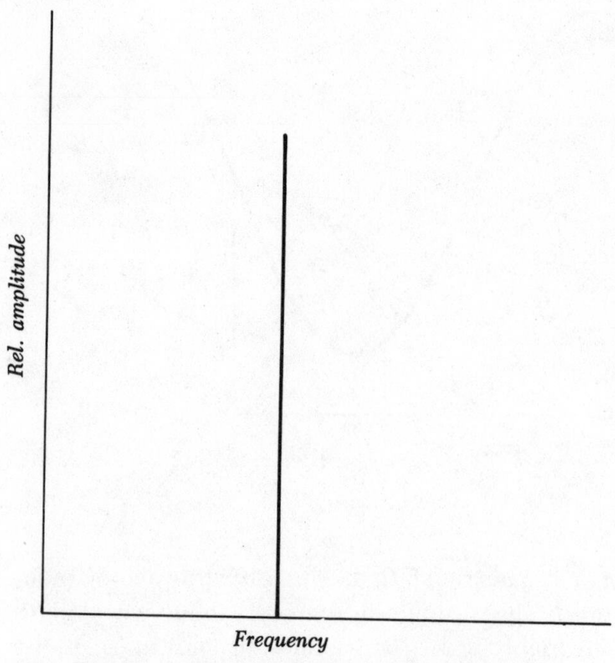

Fig. 1-2. Spectral illustration of a pure sine wave.

has a wavelength of about 7500 angstrom units; an angstrom unit is 10^{-10} meters or 10^{-7} mm. This is too short to visualize—try to imagine one ten-millionth of a millimeter! The shortest wavelength of violet light is about 3900 angstroms, about half that of red light and twice the frequency. There are minor variations among individuals in the range of wavelengths they can see, just as there are variations in the range of sounds we can hear. And other creatures can see different wavelengths than humans. Insects, for example, can see ultraviolet light at wavelengths much less than 3900 angstroms. This is the principle on which ultraviolet bug lamps work. Ultraviolet appears extremely bright to insects.

An incandescent bulb emits light at all wavelengths in the visible spectrum, but most of the energy is concentrated in the red, orange, and yellow regions. A spectral distribution for an incandescent bulb is shown in FIG. 1-4A. Although the light looks almost white, which is in theory an equal amount of light at all visible wavelengths, we can readily see the difference between the light from an incandescent bulb and sunlight. Other kinds of bulbs emit light having different spectral characteristics. A fluroescent bulb is oriented more toward the shorter wavelengths (FIG. 1-4B). Some bulbs emit light at several discrete wavelengths (FIG. 1-4C).

Even those bulbs that emit light at discrete wavelengths, such as the familiar sodium-vapor street lamp, which has two strong emis-

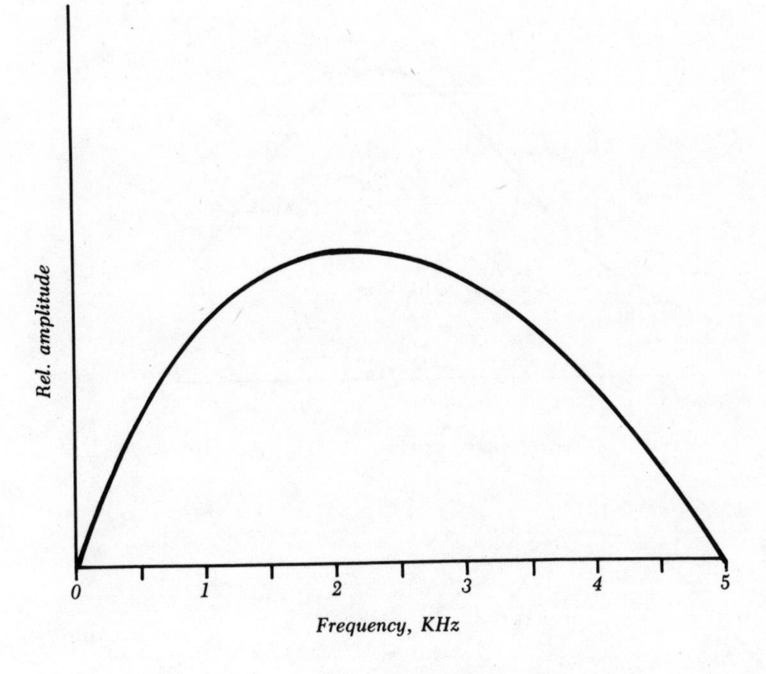

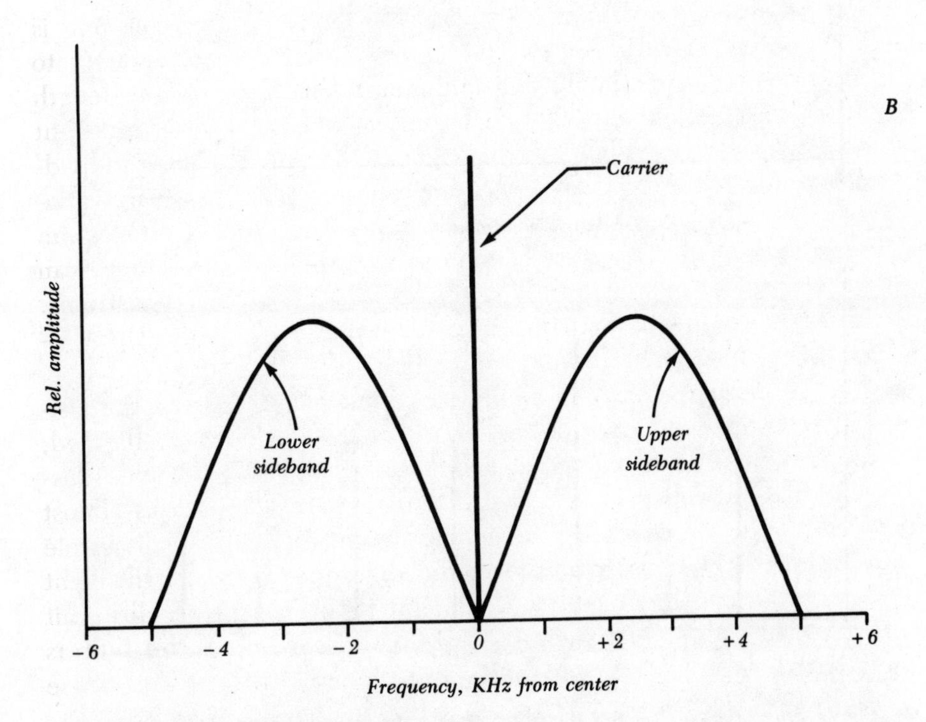

Fig. 1-3. (A) "white noise" at audio frequencies. (B) spectral illustration of a signal modulated with the noise shown at A.

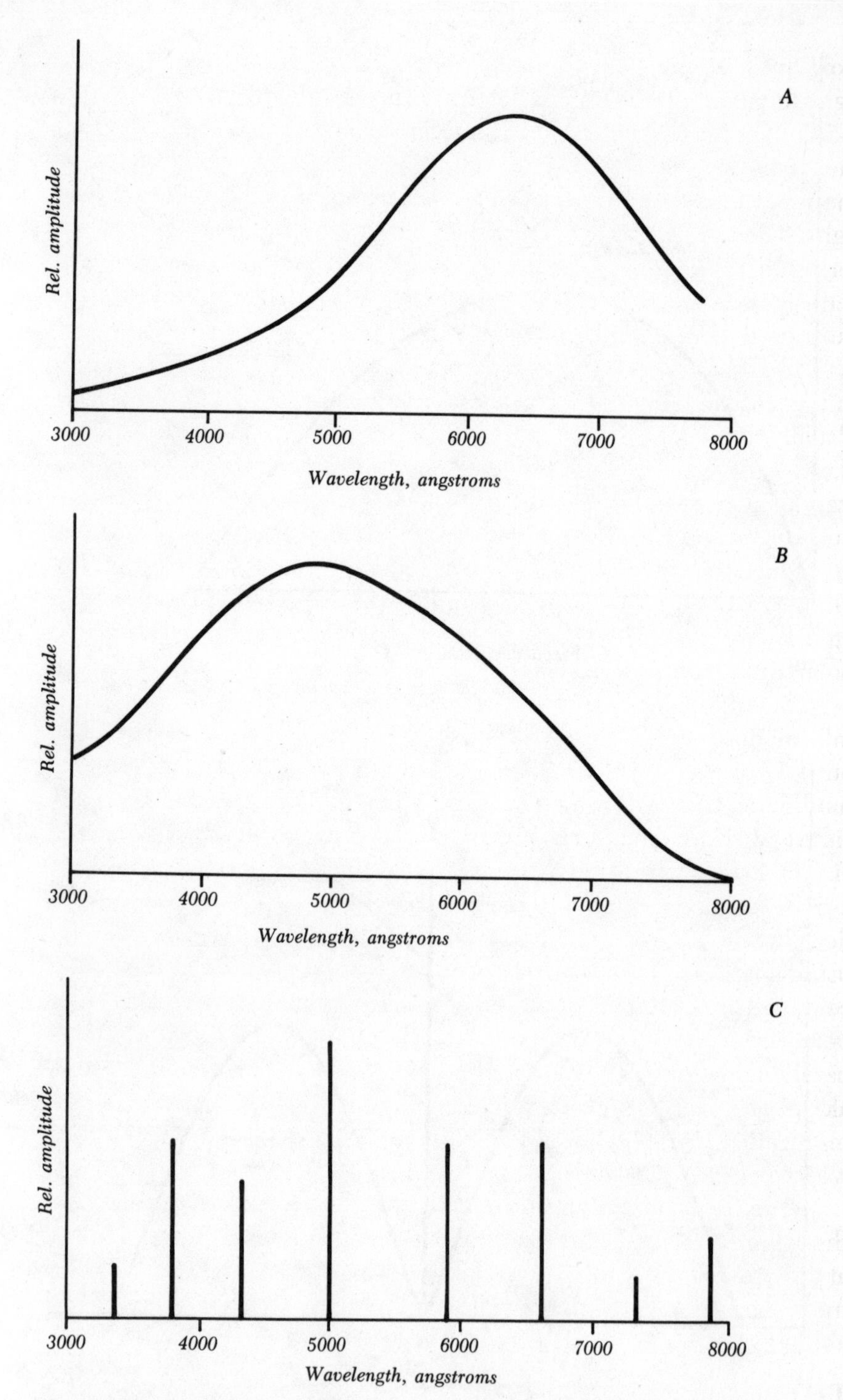

Fig. 1-4. (A) typical spectral distribution of light from an incandescent lamp. (B) typical spectral distribution of light from a fluorescent lamp. (C) hypothetical emission lines from a gas-excitation lamp.

sion lines in the yellow-orange region of the spectrum, emit the light in various phases. The energy is not all aligned in phase (FIG. 1-5A). There still exists a certain *blurriness* in the spectral lines, meaning that the emission is not at a pure, single frequency, although it is much more so than an incandescent bulb or natural daylight. The light is said to be *incoherent* when it has components differing in phase. If the light consists of only one phase component, then it is said to be *coherent* (FIG. 1-5B). No ordinary type of light bulb produces coherent light.

In addition to the light being made up of many different waves in many different phases, ordinary light is polarized in all directions. Usually we speak of polarization as being horizontal, vertical, or slanted. Actually these terms are subjective, as we can define polarization in terms of a 360-degree coordinate circle. This can be reduced to a semicircle if we ignore the phase of the wave. Light can be reduced to a single plane of polarization by means of a polarizing filter, which consists of a transparent material with fine opaque lines etched on it. This is the principle by which polarized sunglasses and polarizing camera filters operate.

Electromagnetic waves can be modulated in many ways to carry information. This is true of both radio waves and light waves. The output of an ordinary incandescent bulb can be amplitude-modulated to a certain extent, simply by placing an audio-frequency signal in series with a DC power source driving the bulb. A simple circuit for accomplishing this, using a flashlight, is shown in FIG. 1-6. A receiver can also be made using a solar cell and audio amplifier (FIG. 1-7). The sensitivity of the receiver can be enhanced by using a large lens or parabolic mirror to focus the light onto the solar cell, and by employing several low-noise amplification stages. The beam from the transmitter can be made narrower by using a large parabolic reflector. With a bright bulb and sufficient modulation, along with large parabolic reflectors and a sensitive receiver, communication can be achieved over distances of several miles in clear weather using this system.

This kind of communications system has the disadvantage that the beam cannot be made very narrow unless a huge mirror is used at the transmitter. Also, energy is scattered over a wide band of frequencies—red all the way through violet, along with considerable infrared—and is in countless different phases.

THE LASER

By employing special methods for generating light, we can obtain a coherent beam of energy with all of the waves in phase, and with an

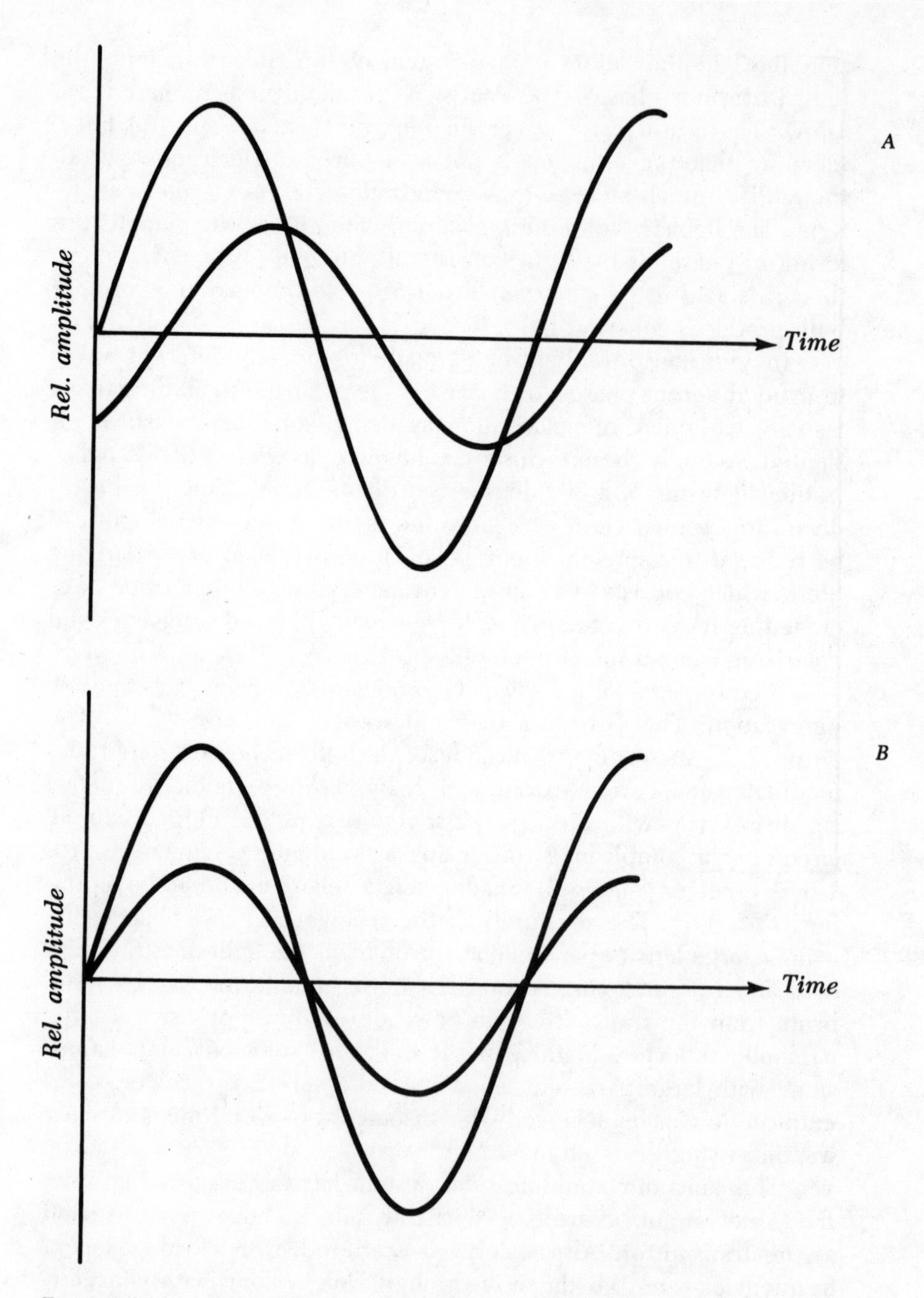

Fig. 1-5. (A) two waves of the same frequency but different phase. At B, two waves of identical frequency and phase. The amplitudes are different.

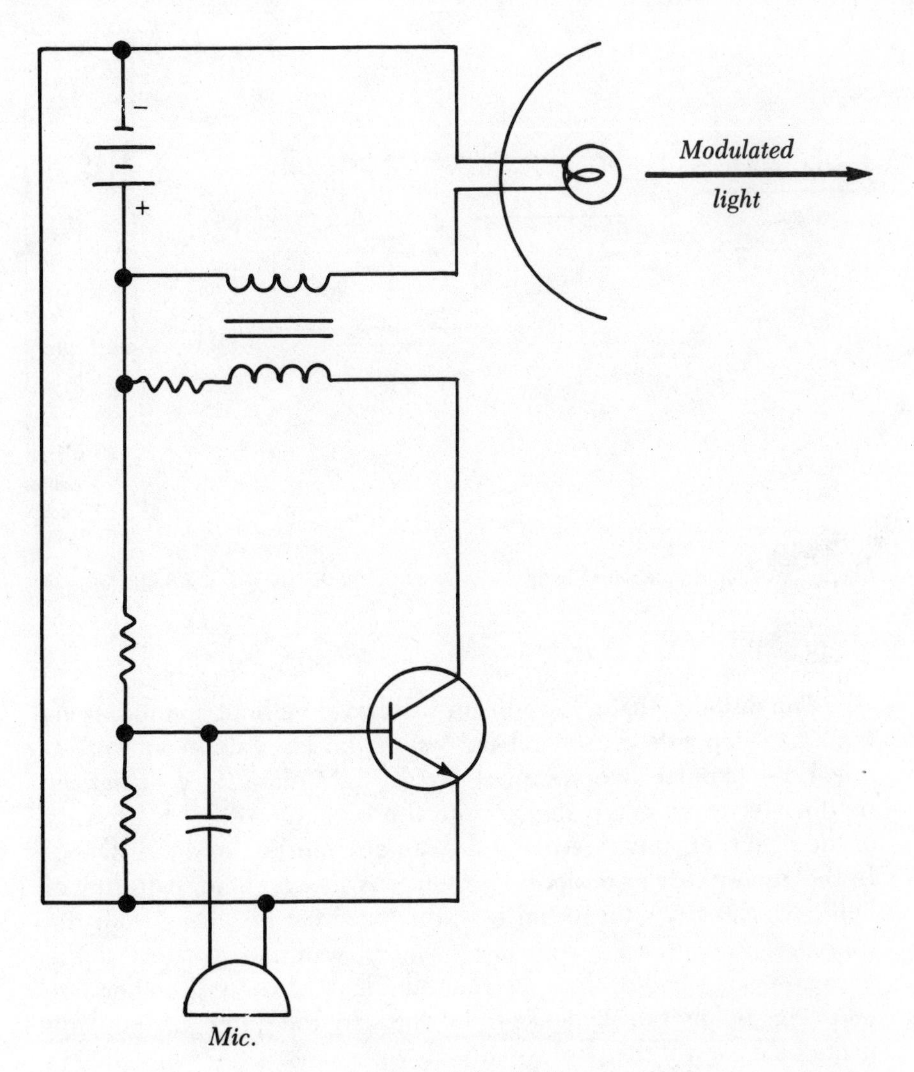

Fig. 1-6. A simple modulated-light transmitter.

extremely narrow bandwidth. All the energy is then concentrated at a single wavelength for all practical purposes and furthermore all the waves are in phase. The result is a light beam similar to a radio-frequency carrier wave.

One of the properties of the laser is that the beam does not spread very much, even over very long distances. The larger the laser in terms of physical size, the narrower, in general, the beam will be. Different kinds of lasers produce beams that spread differently. However, the pencil-thin laser beam is the most adaptable to communications, as well as to other applications, including long-range "ray guns."

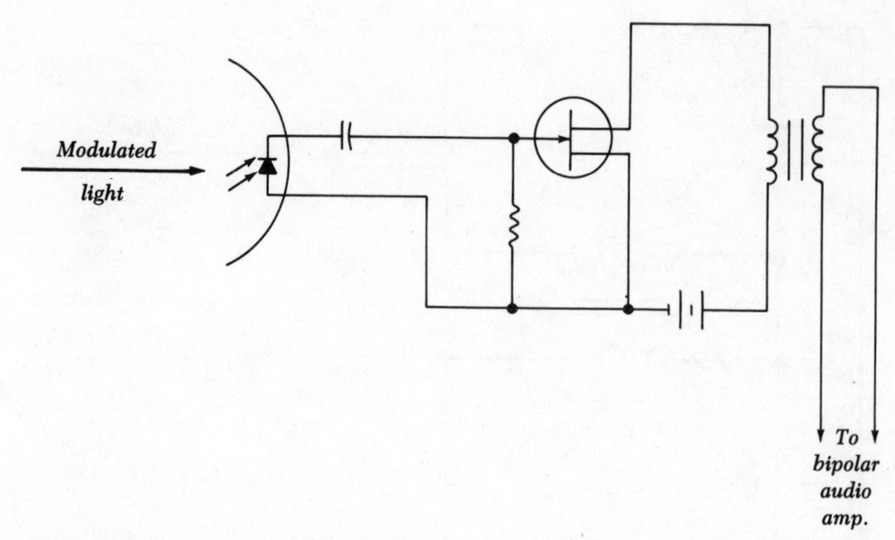

Fig. 1-7. A simple modulated-light receiver. Only the first stage of amplification is shown.

The distance that a laser beam will travel without significant attenuation depends on several factors. If the laser does not spread much, the familiar inverse-square law for light intensity does not apply for a considerable distance from the laser source. The intensity of the beam remains essentially the same regardless of the distance. In the atmosphere or under water, there is always some scattering of light, which causes the beam to spread at great distances from the source. The scattering, both in air and in water, is greatest at the blue and violet end of the spectrum, and least at the red and orange end. This is why the sky looks blue, and why the water in a pool appears blue even if the pool is painted white. When the sun is low in the sky, it attains a yellow, orange, or even red appearance because light at longer wavelengths is more penetrating. So it is not surprising that a red laser, such as the ruby or helium-neon type, has the best propagation in the atmosphere or under water.

In the vacuum of space, the frequency (wavelength) of the laser is not of great importance insofar as scattering is concerned. In this medium, infrared, ultraviolet, and even X-ray lasers become practical, whereas in air, much of the energy at these wavelengths might be absorbed or blocked. We will look at the different types of lasers, the way they work, and the way they are used, throughout this book.

RADIATION AND THE ATOM

An incandescent bulb operates, simply because the filament, carrying a large electric current, gets so hot that it emits visible light as well as infrared light, often mistakenly called "heat". The hotter the filament gets, the shorter the mean wavelength of the emitted radiation. A fluorescent bulb operates on a different principle. The laser resembles the fluorescent lamp much more than it does an incandescent lamp. Excited gases also emit radiation at various discrete wavelengths, and the laser resembles this kind of light source still more. The fluorescent, gas-excitation, and laser light sources all operate on the principle of energy emission caused by electrons changing energy levels within atoms.

In an atom, electrons tend to attain only certain stable energy levels. FIGURE 1-8 is a simplified model of an atom, known as the *Bohr model* after its inventor, Niels Bohr. In fact, the circles shown in FIG. 1-8 are really spheres, and they exist in theory only, but the simplified version shown here is good enough to demonstrate what happens. The larger circles represnt the average instantaneous distance of the electron in a hydrogen atom from the nucleus—a single proton. The energy levels are designated E_1 through E_5. The drawing is not to scale, but the larger circles are higher energy levels. If we denote the relative energy levels for shells E_1 through E_5 by the symbols e_1 through e_5, then $e_1 < e_2 < e_3 < e_4 < e_5$. The electron may "jump" to a higher level or "fall" to a lower level. Clearly, there are many different possible energy level transitions that can take place.

If a photon having just the right amount of energy should strike the electron, the electron will gain just the right amount of energy to "jump" from one energy level to a higher one—for example from e_2 to e_5. This can occur only if the photon has exactly the correct energy content, represented by a definite wavelength. The possible energy transitions may be denoted by means of a table, and they can be numbered, as shown in TABLE 1-1.

Hydrogen has its own set of wavelengths at which it will absorb energy. Other gases and elements have their unique absorption "lines" in the visible spectrum, as well as in the radio, infrared, ultraviolet and X-ray spectra. When radiant energy passes through a gas, the spectrum shows certain dark bands or lines that indicate the presence of a particular element. With visible light, the spectroscope (FIG. 1-9) can be used to identify the presence of various ele-

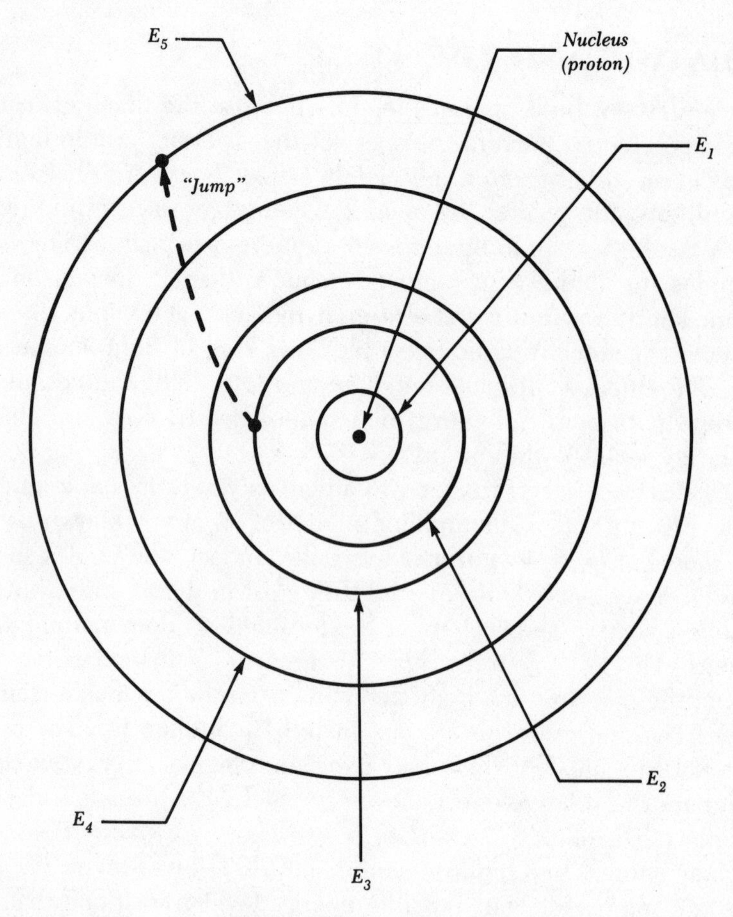

Fig. 1-8. The Bohr model of the hydrogen atom, showing five electron "shells" E_1 through E_5.

Table 1-1. **Designators for energy transitions among five energy levels e_1 through e_5 in an atom. There are ten possible changes in all, in terms of magnitude, designated a through j. These transitions may be either an increase in energy (+) or a decrease (−). Read from the vertical column to the horizontal row, as shown by the example.**

	e_1	e_2	e_3	e_4	e_5
e_1	0	+a	+b	+c	+d
e_2	−a	0	+e	+f	+g
e_3	−b	−e	0	+h	+i
e_4	−c	−f	−h	0	+j
e_5	−d	−g	−i	−j	0

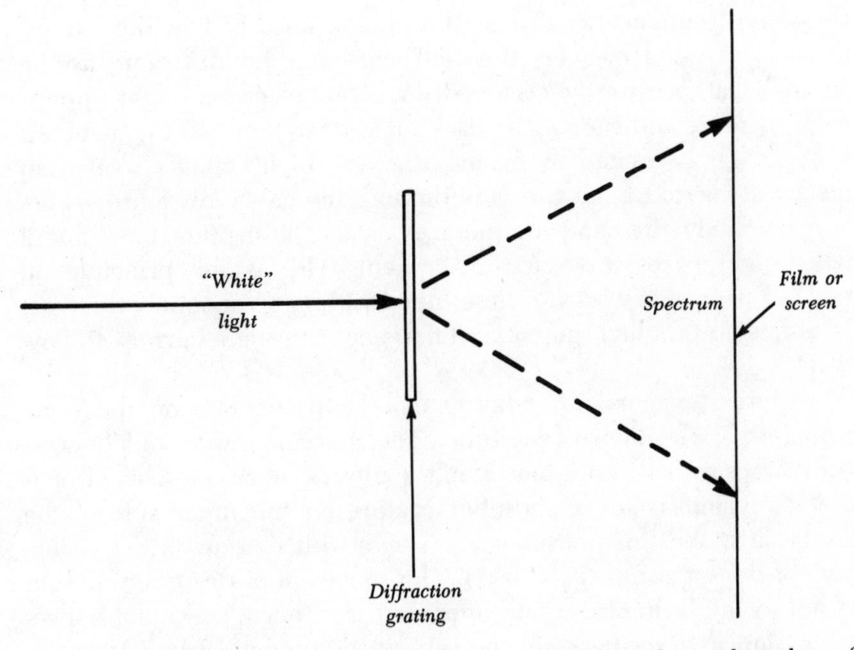

Fig. 1-9. Simplified diagram of a spectroscope. A prism may be used in place of the diffraction grating.

ments. This instrument is used by astronomers to observe distant stars and find out what gases are within the star, or between us and that star. The spectrum of the Sun, for example, is very complicated, indicating the presence of many different elements in our parent star. Whole galaxies have been observed and their absorption lines recorded. Distant galaxies show the same pattern of lines as nearer ones, but the lines of more distant galaxies are all displaced toward the longer (redder) wavelengths. This so-called *red shift* was noticed early in this century and was explained by hypothesizing that distant galaxies are moving rapidly away from us—so fast, in fact, that a phenomenon called the *Doppler effect* causes a change in the wavelength of energy emitted by the stars in those galaxies as the energy arrives here on Earth. This observation ultimately led to the contemporary theory of an expanding universe.

Just as gases will absorb energy at certain wavelengths as electrons "jump" from lower to higher energy levels, the opposite also can happen. When a gas absorbs a large amount of energy, the electrons tend to "fall" back to lower energy levels, going around the nucleus in smaller shells. An element seems to be more "satisfied" at a lower energy level as compared to a higher one. When an electron "falls" from, say, energy level e_5 to e_2, a photon is emitted at exactly the same frequency as that of the photon absorbed in the "jump" from e_2 to e_5. However, the "fall" may not be the same as the "jump" that previously occurred. A given electron might "jump" from e_2 to e_5 and then "fall" back, at a later time, to e_1, e_3 or e_4. When a gas is excited by means of a very high voltage, so that an electrical current begins to flow through the gas, emission lines occur at exactly the same frequencies as the absorption lines found when light pases through the element. This is the principle on which a neon sign works. A tube, filled with neon or some other gas, is subjected to a high potential difference, causing a current to flow (FIG. 1-10).

Mercury-vapor and sodium-vapor lamps operate on the same principle as the fluorescent tube. The mercury-vapor and fluorescent lamps actually emit mostly in the ultraviolet range. This ultraviolet emission strikes a phosphor coating on the inner side of the lamp glass, and this phosphor gives off a whitish glow in the visible part of the spectrum (FIG. 1-11). The spectrum of this emitted light is not in the form of discrete lines, but is rather a continuous spectrum similar to daylight. In the case of the mercury-vapor lamp especially, significant ultraviolet light does escape, and recently there has been some concern that this light might be harmful to people exposed to it for long periods at high intensity. The familiar "sunlamp" uses mercury vapor to produce these rays that give us an indoor suntan, and also possibly a dangerous burn.

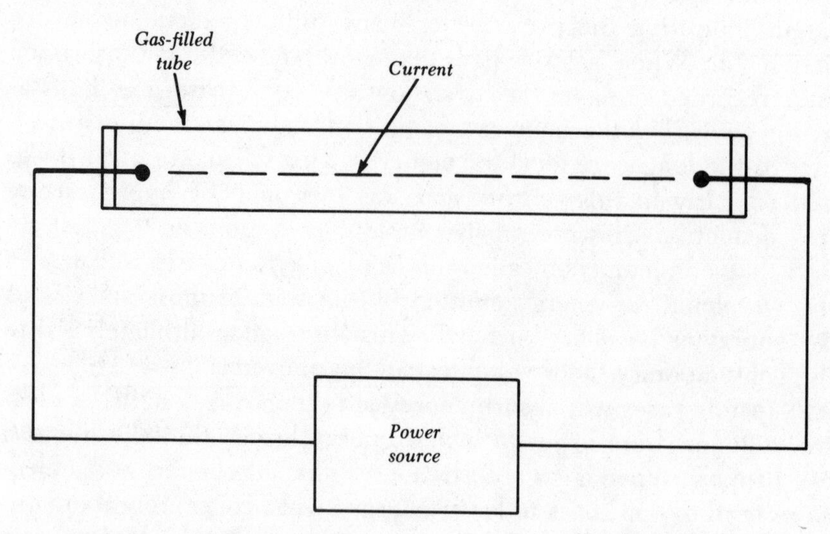

Fig. 1-10. Simplified diagram of the operation of a gas-excitation tube, such as a neon-sign tube.

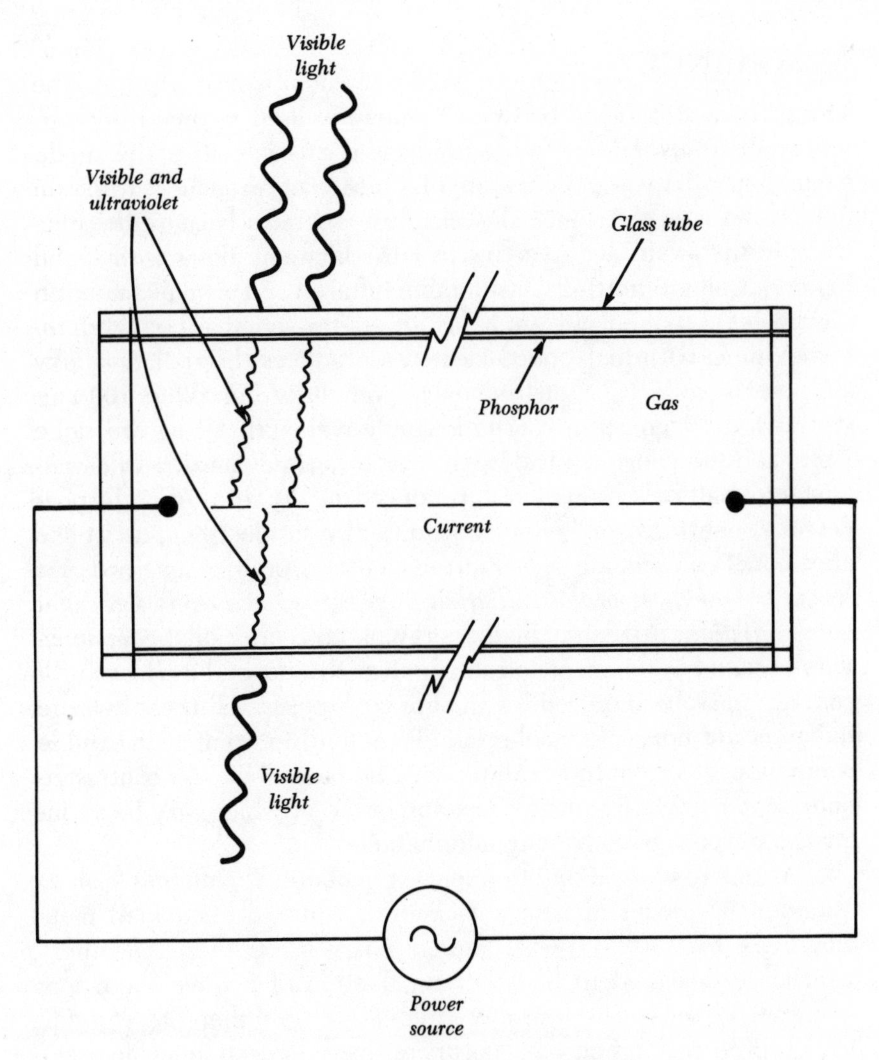

Fig. 1-11. Principle of the fluorescent tube.

The quantum principle helps us to understand how the laser operates. When a beam of light passes through a substance in which the majority of the atoms are in a high-energy state, the beam tends to be amplified at certain wavelengths. This is called *stimulated emission*. The amplified beam is a coherent beam, in which all of the wavefronts are in phase. If the atoms are in a sufficiently high energy state, and if the beam of light passes through the substance many times, an extremely intense beam of coherent light is produced. This is the basic principle by which the laser works.

RESONANCE

The primary difference between a source of light in the ordinary or more often-encountered sense, and radio waves, is that the single-frequency radio wave is generated by means of resonance, while the light beam is not. An incandescent bulb operates because the electrons in the atoms are so active that the filament glows with visible light, as well as emitting considerable infrared and a small amount of ultraviolet radiation. In an X-ray tube, the bombardment of the heavy metal with high-speed electrons produces the radiation, usually over a wide band of frequencies ranging from about 100 angstroms to 0.01 angstroms. (The longer-wavelength X-rays are called "soft" and the shorter-wavelength rays are called "hard.") In certain unstable materials, emission is produced in the form of high-speed protons, neutrons, or helium nuclei, and also electrons, as well as photons. The radiation of helium nuclei (two protons and two neutrons) is known as *alpha radiation*; high-speed electrons are called *beta radiation*; extremely high-energy photons are called *gamma radiation*, and have such penetrating power that several inches of solid concrete may be required for shielding purposes. All these forms of radiation are normally incoherent. They tend to occur with random components of amplitude and phase. The radio wave, by contrast, is energy at a single frequency, the amplitude of which may be varied for the purpose of conveying information.

At microwave radio frequencies, coherent radiation can be obtained by means of resonant cavities, which are shielded metal chambers that have special dimensions. Inside these chambers, which are designed to be a certain fraction of a wavelength long (generally ½ wavelength or any integral multiple thereof), *standing waves* form. A standing wave occurs as the radio-frequency energy is propagated back and forth in the chamber, reinforcing itself in phase each time it traverses the chamber. The result is reinforcement, or amplification, of the signal. This closely approximates the way lasers work. A resonant circuit causes an amplifying electronic circuit to oscillate, or generate coherent signals at certain frequencies.

Within a resonant cavity, the magnetic (M) field is always perpendicular to the electric (E) field. (Note: The M field is often referred to as the H field). When this condition exists, an electromagnetic (EM) field is formed. The EM field has the property of being radiated great distances through space. This EM field can be transferred to an antenna, such as a horn, via a long, rectangular chamber known as a waveguide (FIG. 1-12). The waveguide is designed so that it is resonant in terms of the cross-sectional size.

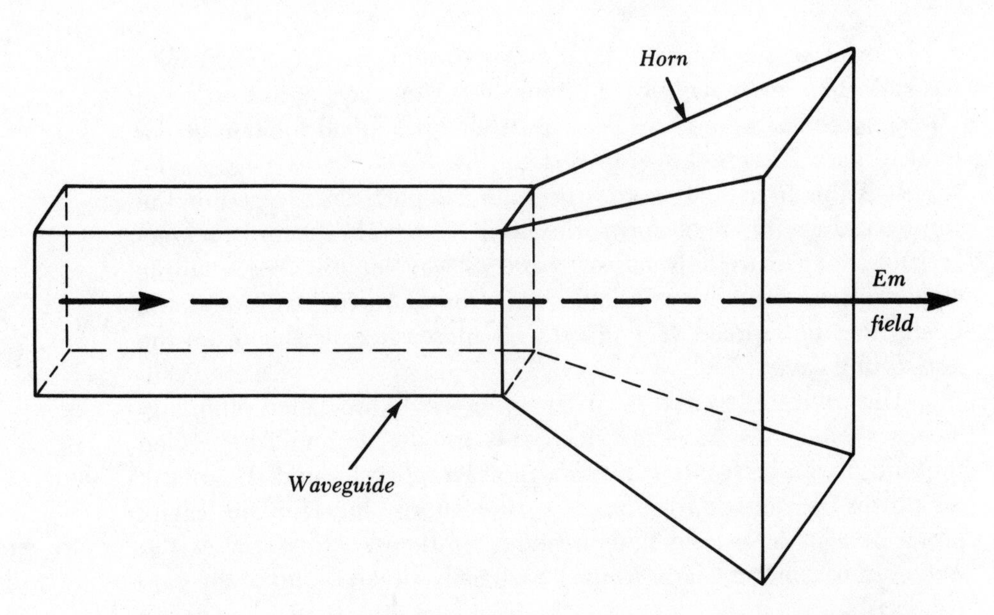

Fig. 1-12. A method of transmitting microwave energy from a waveguide.

The resonant cavity is *excited*, or fed a signal, by means of a probe, loop, or cylinder. The probe introduces an electric field into the cavity. The changing electric field produces a fluctuating magnetic field at right angles to the E field. The result is electromagnetic resonance, provided that the cavity is the right size. A coil introduces the magnetic field into the cavity. The changes in the M field produce an E field at right angles to the lines of flux in the M field, and again, provided the cavity is of the correct dimensions, resonance occurs. The cylinder can be used to introduce clusters of high-speed electrons into the cavity. These electrons actually constitute an electromagnetic field themselves, and if they are of a frequency that corresponds to the dimensions of the cavity, resonance again takes place. This last method of exciting a resonant cavity at microwave frequencies most closely approximates the true optical laser.

The heart of most lasers consists of a cavity that is resonant at visible wavelengths. Since the wavelengths of visible light are extremely short, optical cavities are many wavelengths long, otherwise they would be of prohibitively small size. The atoms in the cavity, whether they be gas, liquid, or solid, are excited so that their electrons are raised to high energy levels. This causes them to emit light at certain frequencies. The cavity is of such a size that it is resonant at one of these emission frequencies. The result is amplification of the energy at this frequency, and generation of coherent light.

How does the coherent light escape from the cavity? The cavity has reflective inner surfaces at either end. However, at one end, the mirror is not completely reflective; it allows a small fraction of the light to pass. This is shown in FIG. 1-13.

It is this light that escapes through the partially silvered end of the optical cavity which forms the laser beam. The beam is characterized by an extremely narrow band of wavelengths, essentially a single optical frequency. All the wavefronts are in phase, and the beam does not spread very much with increasing distance from the generating cavity.

The cavity is excited by means of a technique called *pumping*. There are various ways in which this can be accomplished. One method uses a high-intensity flash tube wrapped around the cavity. As pulses of excitation are applied, the energy levels of the cavity atoms reach higher and higher values, until *breakthrough*, or the emission of coherent light from the partially silvered end of the cavity occurs.

The orientation of the silvered ends of the cavity is critical: they must be perfectly parallel, or as close to parallel as possible. If the ends are not parallel, the light beam will not be reflected back and forth very long before striking one of the cavity walls.

The earliest maser used vaporized ammonia in its cavity. This device produced a narrow, coherent beam at extremely high radio frequencies. More recently, optical masers (lasers) were developed using different gases, liquids or solids in their cavities.

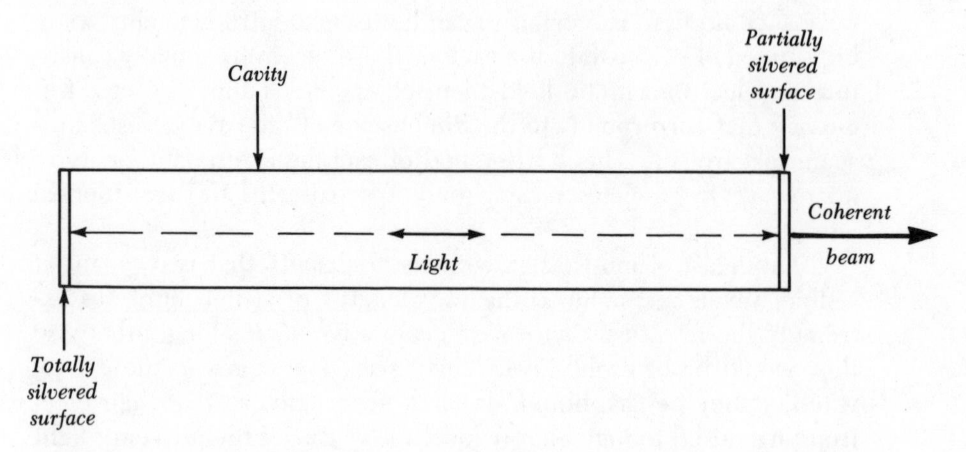

Fig. 1-13. Simplified illustration of the principle of the laser. The pumping apparatus is not shown.

HOW LASERS AMPLIFY

To clarify the amplification process—the means by which a laser actually magnifies the intensity of the light in its cavity—we must be aware of the fact that during an energy transition in an atom the impinging particle is not absorbed by the atom. Instead, it continues on in the same direction and with the same wavelength (energy level) as before.

An example of the amplification process is shown in FIG. 1-14. A photon, having a certain wavelength, arrives at an atom and causes an electron to drop to a lower energy level. The transition occurs such that a new photon is emitted, going in the same direction and having the same wavelength as the original photon. The original photon, P_1, must have just the right energy level for resonance with the energy transition within the atom, so that photon P_2 has the same wavelength as P_1.

Photon P_1 is not absorbed, but continues along with P_2. The interaction is somewhat akin to passing of a baton in a relay race, with the finishing runner continuing on beside the last runner in-

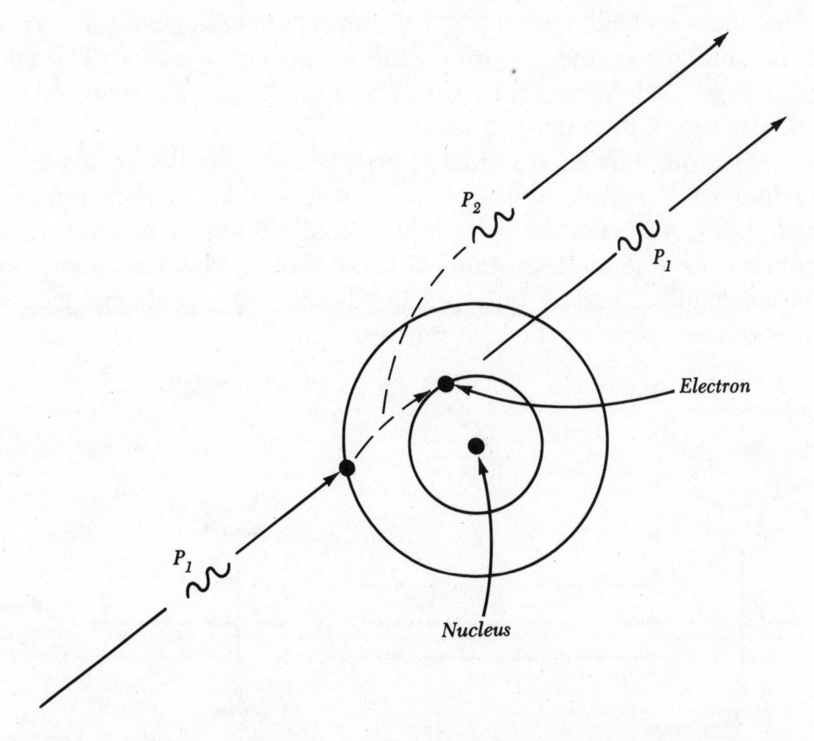

Fig. 1-14. Amplification of laser energy by means of a photon interacting with an atom.

stead of stopping. The two waves are in phase because of the nature of the collision; therefore, amplification of the wave has taken place. This cycle occurs many times as the pumping source continues to restore the atoms to a highly excited state and the photons are reflected back and forth within the cavity. The same atom may be responsible for many photon emissions. Of course there are billions of atoms in the cavity, and the reflection occurs many thousands of times. If the cavity is just the right length, that is, an integral multiple of the resonant frequency for the beam of light (or other electromagnetic energy), then amplification can take place to an extreme degree. Even a moderately intense laser will produce light many times more intense than direct sunlight. For this reason lasers have acquired the reputation of being "ray guns" capable of burning holes in metal or concrete barriers, or shooting down aircraft and satellites. It is true that high-intensity lasers can do this, but these examples represent only a small portion of the applications for these devices.

All lasers are characterized by: (a) a cavity, usually in the shape of a cylinder or prism; (b) resonance, determined by reflectors; (c) a source of energy, called the pumping source. The medium determines the wavelength at which the laser will operate, both in terms of the substance(s) employed and the length of the cavity. The energy output is determined by the size of the laser and especially by the intensity of the pumping source.

Since the mirrors are exactly, or as closely parallel as possible, maximum intensity is in a direction perpendicular to both mirrors (FIG. 1-15). Although emission occurs in all directions as a result of spontaneous energy transitions caused directly by the pumping source, amplification is in a parallel beam that eventually passes through the partially reflective mirror.

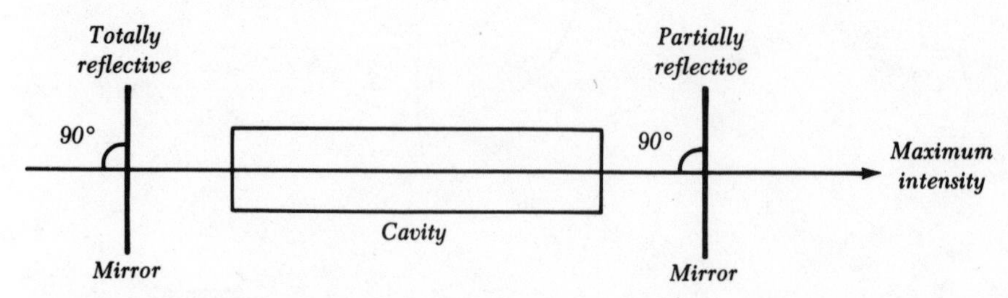

Fig. 1-15. Maximum beam intensity occurs in a direction perpendicular to both mirrors.

FOCUSING OF LASER ENERGY

Laser beams are focused in the same manner as ordinary light. Since light from a distant noncoherent source is essentially parallel, it can be brought to a sharp focus and will cover a very small area. The size of the spot is determined by the distance of the source, the size of the source, and the focal length of the lens or mirror. We are all familiar with the way sunlight can be focused to cause a piece of paper to catch fire. The larger the area of the lens or mirror, the more light is gathered and the more intense the spot, assuming constant focal length.

For a laser, the situation is somewhat different. The area of the lens or mirror need only be enough so that the whole beam is captured. This is, in theory, the same no matter what the distance of the laser from the lens (FIG. 1-16A). In practice there is some spreading of the beam, but it is not significant unless the distance from the laser is very great (FIG. 1-16B).

Assuming a parallel beam, which we can for all practical purposes in the case of a laser, it is possible to obtain an extremely small spot in the focus. Theoretically, this spot is a geometric point no matter what the focal length of the lens or mirror. In practice it is not a perfect point. It would have infinite energy concentration if it were, and this energy would be concentrated in an infinitely tiny area—a paradox. Instead it is an extremely intense spot of miniscule

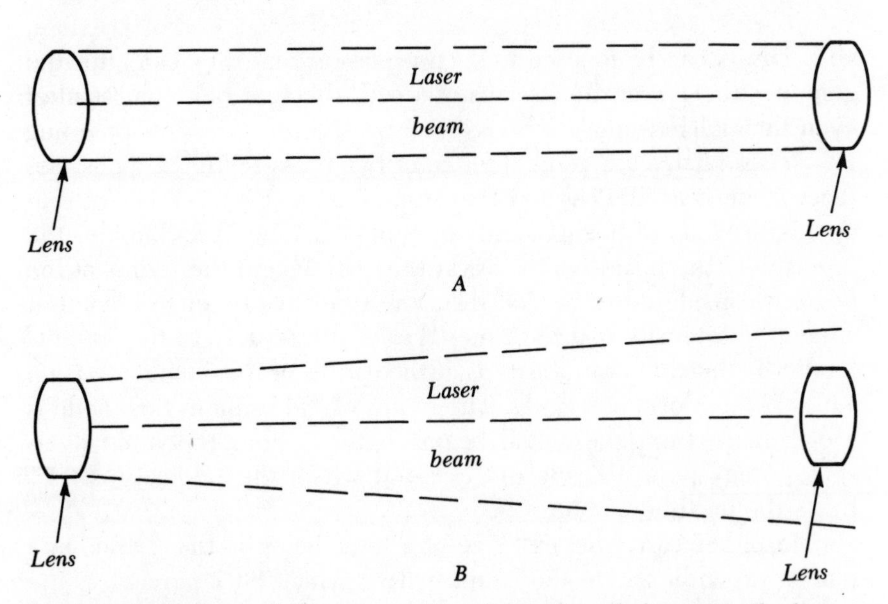

Fig. 1-16. Perfectly parallel laser beam (A) and spreading of a beam (B). The former is a theoretical case.

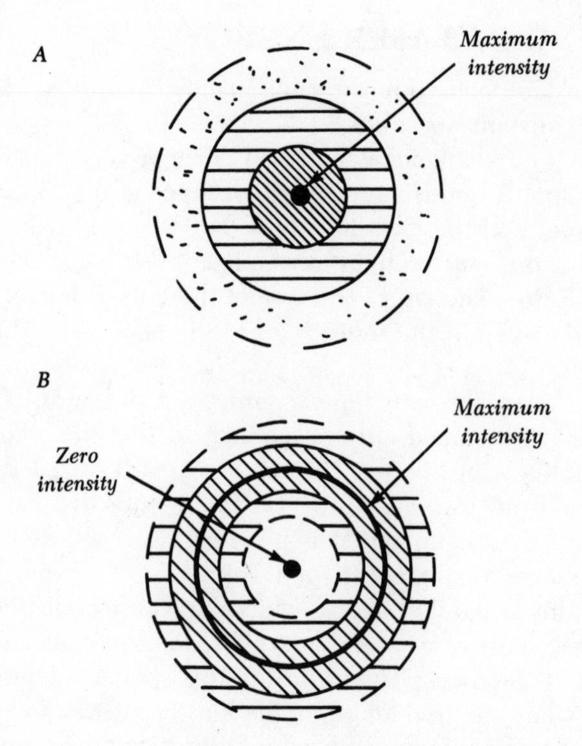

Fig. 1-17. The "spot" laser mode (A) and the "donut" mode (B). Both of these show cross sections in a plane perpendicular to the axis of the beam.

size. Lasers can be focused to such precision that they can alter the genetic structure of the nucleus of a cell, the spot being far smaller even than a bacterium.

Laser beams are generally one of two cross-sectional types: the "spot" type (FIG. 1-17A) and the "donut" type (FIG. 1-17B). In theory, both kinds of beams focus to a point. In the "spot" mode, the intensity of the beam is greatest at the center, and the radius of the beam is considered to be that distance from the center to a point at which the intensity is 0.133 times that of the center. In the "donut" mode, the maximum intensity is at the center of the "donut" portion and takes the form of a circle. The radius of the beam in this mode is the distance from the actual beam center, where the intensity is zero, to any point outside the circle at which the intensity is 0.27 times the maximum value.

In either case, the spot size of a laser beam is that radius in a plane perpendicular to the beam, within which 86.7 percent of the light is transmitted. This radius is shown for the "spot" mode in FIG. 1-18A and for the "donut" mode in FIG. 1-18B.

Laser energy can be obtained by using gaseous, liquid or solid media, and sometimes by using combinations of matter in two different states. We will now examine some of the more common kinds of lasers and see how they are built.

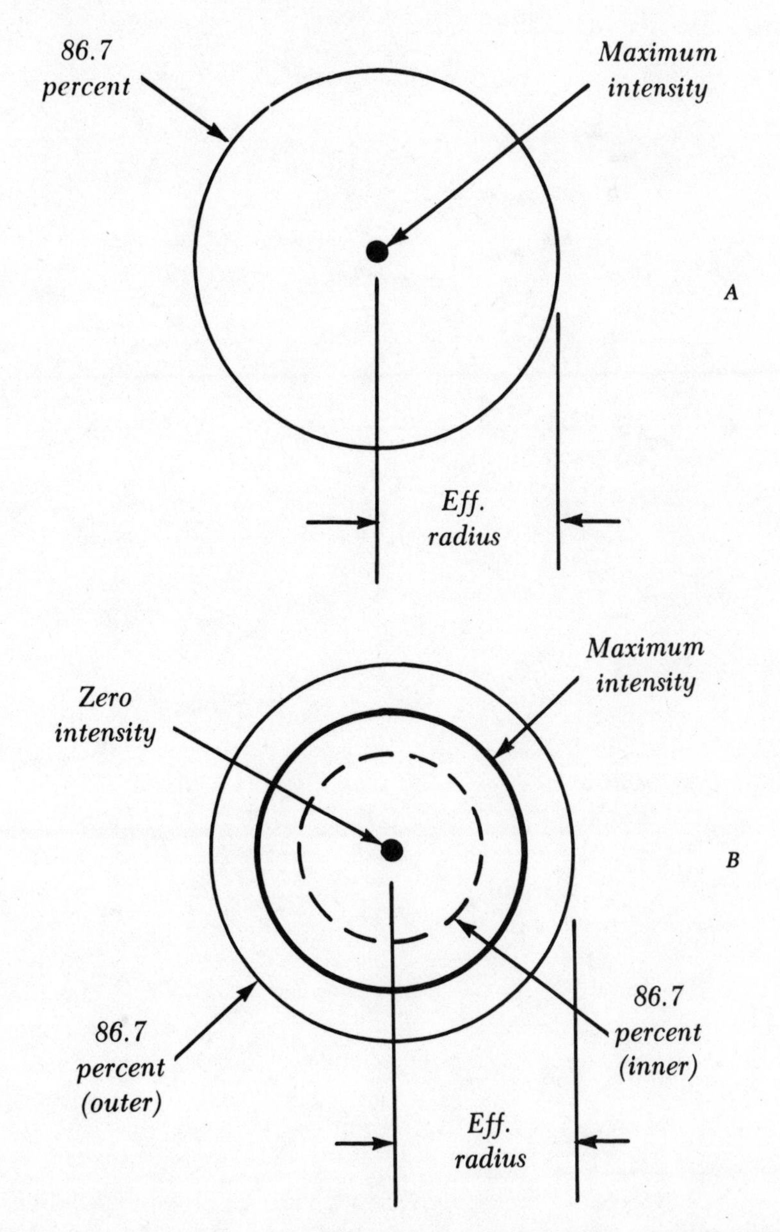

Fig. 1-18. The radius of the beam is considered to be that radius within which 86 percent of the energy lies, in a cross-sectional plane perpendicular to the axis of the beam. (A) "spot" mode; (B) "donut" mode.

2

Types of Lasers

Coherent light can be generated in many different ways. Lasers vary in how they are excited (pumped), in the materials used to obtain the necessary resonance, in the power generated and consumed, in wavelength, and in terms of usefulness for various applications. An especially important classification of laser is the solid-state laser, exemplified by the laser diode.

Not all lasers are capable of burning holes through solid concrete or steel walls and doors, or of knocking aircraft or satellites out of commission. But some lasers can do this, and we already have the technology to construct and use laser weapons. This makes the laser a powerful tool indeed, and the controversy surrounding the development of high-intensity lasers arises from this potential for destruction. Powerful lasers need not be used as weapons; however, they can be used for welding in the vacuum of space, for example. They can also be used to measure long distances, such as the separation between Earth and the Moon to within a fraction of a centimeter. Laser communications may someday be established between the Earth and distant colonies on other planets in the Solar System.

BEHAVIOR OF SEMICONDUCTORS

A complete discussion of semiconductors and their characteristics is not given here, but some understanding of how semiconductor materials are used in modern electronic circuits is essential to the appreciation of the operation of a semiconductor laser. For more information about semiconductors and their characteristics, refer to *Basic Transistor Course, 2nd Edition*, TAB BOOKS, Inc., #1605, or an equivalent elementary book about solid-state electronics.

Semiconductor materials are, as the term implies, partially conducting. Their conductivity can be varied, depending on the direction in which the current flows, and in the case of bipolar and field-effect transistors, on the bias applied to a semiconductor junction. There are two kinds of semiconductor material, N type and P type. In the N-type semiconductor, the current flows by means of the transfer of negative charge carriers (electrons) from atom to atom, just as in ordinary copper wire. In the P-type semiconductor, the charge carriers are positive, represented by the transfer of electron shortages known as "holes." The physicist always considers the direction of current flow to be from positive to negative; thus, while the concept of hole transfer is more difficult to grasp intuitively, it actually represents a more consistent view with regard to the physicist's theoretical picture of current flow. Holes move "forward" and electrons "backward" with respect to the direction of current.

The simplest semiconductor device is the diode, or two-element device. Two pieces of semiconductor material are placed in contact, one being of the P type and the other of the N type. Alternatively, a fine wire is put in contact with a piece of P type material. The result is called a *P-N junction*. This junction behaves in a certain manner when a voltage is applied in one polarity, and in a vastly different way when the polarity of the voltage is reversed. If the positive pole is applied to the P-type material and the negative voltage to the N-type semiconductor, current flows through the device provided the voltage is more than the "breakover" voltage. This voltage is small—on the order of a fraction of a volt. In the opposite case, where the P-type material receives a negative charge and the N-type material a positive charge, practically no current flows unless the voltage is quite high. This value may be tens or even hundreds of volts, and is known as the "avalanche" or "breakdown" voltage. Typical semiconductor diode current-versus-voltage curves are shown for the forward-bias case (FIG. 2-1A) and for the reverse-bias case (FIG. 2-1B).

The behavior of the semiconductor diode makes it useful as a rectifier, and if the capacitance at the junction in the reverse direction is small, diodes can be used as detectors for radio-frequency signals. Diodes are also used in many other ways, depending on their characteristics: they can be employed as mixers, voltage limiters, and even oscillators in the UHF and microwave radio-frequency

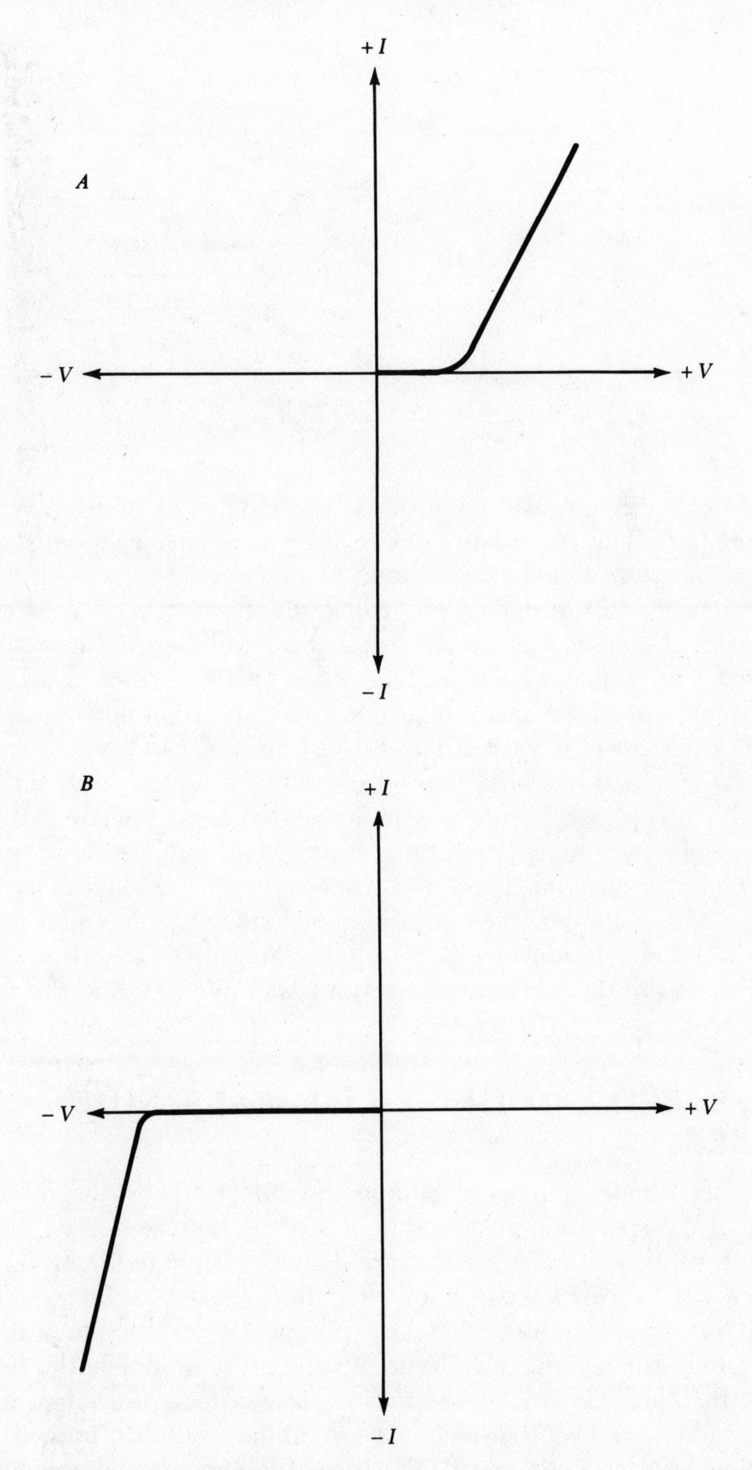

Fig. 2-1. General curves for semiconductor diodes. (A) forward bias; (B) reverse bias.

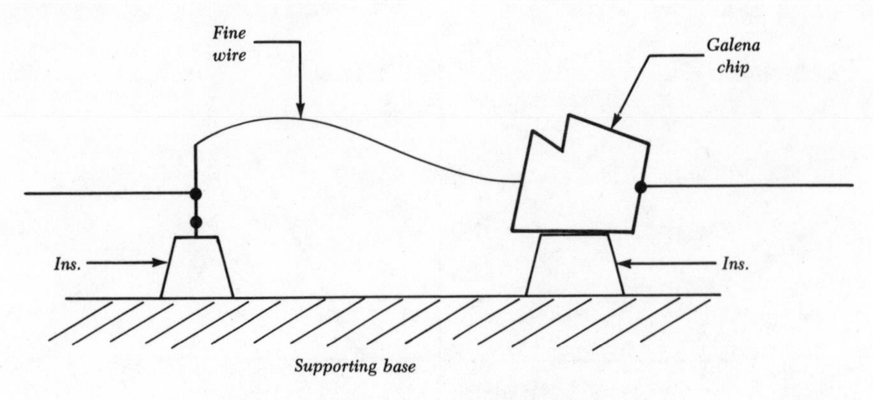

Fig. 2-2. Early galena-chip "cat's whisker" diode.

bands. Diodes can also be used to generate light or to generate power from light. A certain kind of diode generates coherent light, and is therefore called a laser diode.

Semiconductor diodes are fabricated from a variety of substances. The earliest diode consisted of a small chip of the mineral galena with a fine wire placed in contact with it (FIG. 2-2). This "cat's whisker" had a small capacitance at its junction because of the small diameter of the wire. The earliest form of radio receiver, the crystal set, used this technique to demodulate signals. You can still build a crystal set in this way and receive strong signals with no source of power except the signal itself. More modern devices use silicon or germanium, treated in a special way, to obtain diode action. Other materials, such as gallium arsenide are sometimes used. This substance produces coherent light when it is properly treated and when the right amount of current passes through it in the right direction.

INFRARED AND THE GALLIUM ARSENIDE LASER

The laser diode employing gallium arsenide works in the infrared part of the spectrum, so, we cannot see the emissions. The wavelength is just a little longer than the longest visible red light, and is called *near infrared*. Longer wavelengths are known as *intermediate* and *far infrared*. Infrared light is important for communications purposes. The fact that it is invisible makes it more difficult to intercept than ordinary visible light. Guidance systems use infrared lasers and infrared is extensively used in military systems. Infrared radiation travels at the speed of light in a vacuum, and somewhat more slowly in infrared-transparent media other than a vacuum. The near infrared can be focused and refracted just like ordinary light. Red light extends down to a wavelength of about 0.75 microns, or 7500 angstroms. Infrared extends down to about 1000 microns, and is classified as *near infrared* (0.75-1.5 microns), *intermediate infrared* (1.5-5.6 microns) and *far infrared* (5.6-1000 microns). Below 1000 microns, the radiation is generally classified as microwave. The infrared spectrum, with visible light at shorter wavelengths and microwaves at longer wavelengths, is illustrated in FIG. 2-3.

Infrared is not literally heat, although it is sometimes called "heat radiation." When infrared radiation strikes an object that absorbs it, such as the skin, heat is produced. This is why we feel heat in the presence of infrared. The heat from infrared is quite penetrating in animal tissues and it is therefore extensively used in the treatment of pulled muscles and tendons. All objects emit some infrared radiation, regardless of their actual temperature. The hotter an object, in general, the more electromagnetic radiation it emits. Hotter objects emit more energy at shorter wavelengths than cooler objects. This is why hotter stars look bluish while cooler ones appear orange or red.

Infrared systems are fairly simple to construct, are inexpensive, and have good definition and resolution. A satellite guidance system developed by NASA uses infrared devices in which gallium arsenide lasers are used for tracking and rendezvous. This infrared system works better than radar because the wavelength is much shorter, offering superior resolution, especially at short ranges.

The emissivity, or degree with which an object absorbs or emits infrared radiation, is important in the use of infrared laser devices. Some materials such as polished metal have poor emissivity. Painting an object flat black will improve its emissivity. The emissivity may be expressed as a percentage, or as a number from 0 to 1. An emissivity figure of 0 means that the object theoretically emits no energy at a particular wavelength, while an emissivity of 1 means it emits all the energy at a certain wavelength. Clearly, emissivity depends on the wavelength. For example, a red ball has good emis-

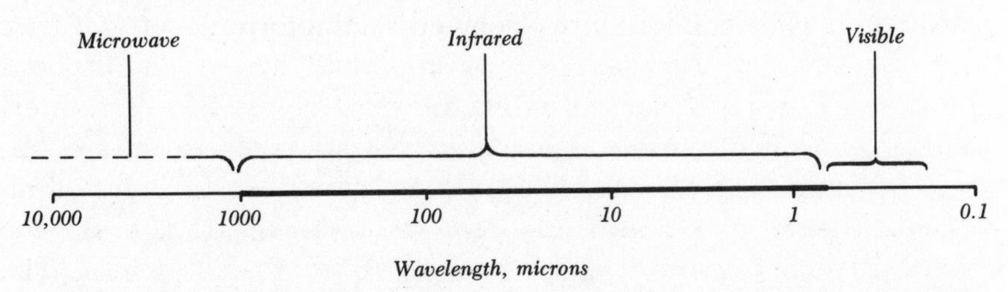

Fig. 2-3. The infrared spectrum lies between microwave radio and visible light.

sivity at wavelengths around 4000 angstroms but poor emissivity around 7000 angstroms.* The best emissivity is obtained by an object known as a black body. This is a theoretical object that absorbs and emits energy with 100-percent efficiency at all wavelengths from radio to gamma rays. The emissivity is the ratio of the energy emitted at a certain temperature by an object, as compared with the energy that would be radiated at the same temperature by a black body.

The hotter an object, the higher the frequency at which most of the energy is radiated. The amplitude-versus-wavelength graph for any object is shown generally by FIG. 2-4. The curve is characterized by a peak, with lower amplitudes at wavelengths longer and shorter than the maximum. The peak amplitude is easy to determine and the wavelength at which it occurs is also easy to find. The exact wavelength at which the peak radiated intensity occurs can be found according to the formula

$$\lambda = \frac{2900}{T}$$

where λ is the peak wavelength in microns and T is the temperature in degrees Kelvin. The Kelvin temperature scale is also called the absolute temperature scale, because it has a value of zero for the coldest possible temperature. To convert degrees Celsius into degrees Kelvin, subtract from 273:

$$K = 273 - C,$$

where K is the absolute temperature in degrees Kelvin and C is the temperature in degrees Celsius. If F is the temperature in degrees Fahrenheit, then

$$K = 273 - [\frac{5}{9}(F - 32)]$$

The overall intensity of the radiation—the area under the curve in the graph of FIG. 2-4—increases as the temperature of an object gets higher. The actual intensity is given by the formula

$$e = T^4ES,$$

*In general, emissivity is poor at wavelengths where reflectivity is good, and vice versa.

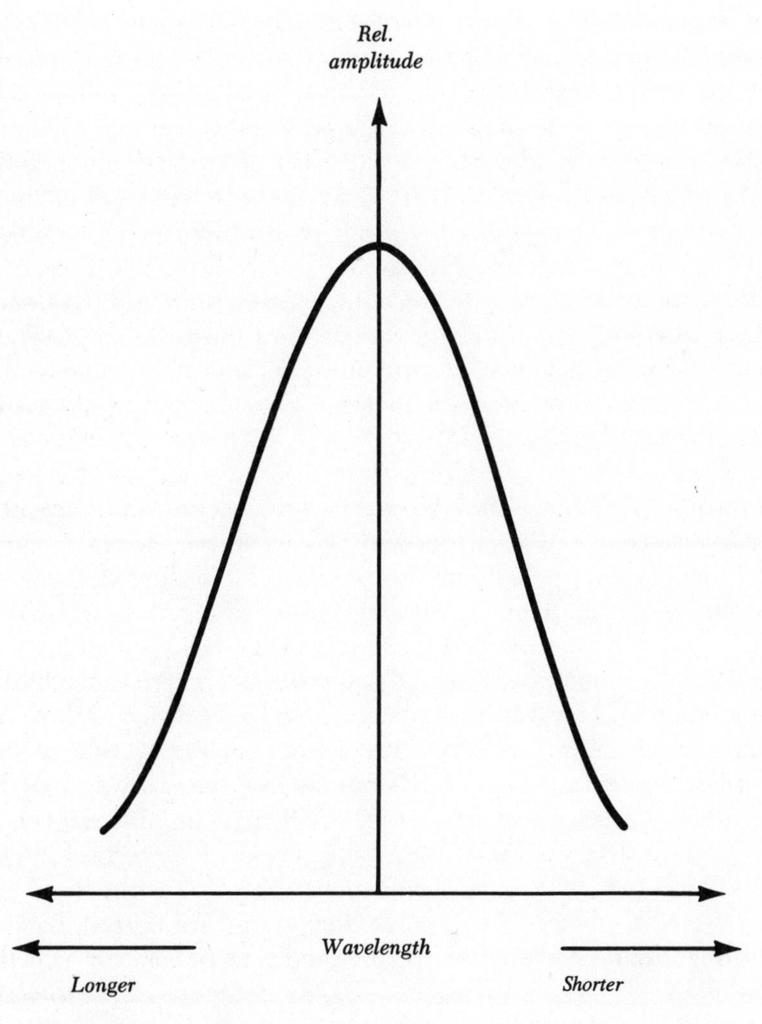

Fig. 2-4. Typical emission curve for an object at any given temperature.

where e is the energy in terms of power per unit area, T is the absolute temperature in degrees Kelvin, S is a proportionality constant and E is the emissivity factor (between 0 and 1).

The atmosphere is not perfectly transparent to all visible wavelengths, and the same holds true for infrared. Certain wavelengths of infrared are severly attenuated by the earth's atmosphere. We would not want to use these wavelengths for long-distance communication, guidance, or any other application requiring that the instruments be separated by a medium of air. Dust in the air, and also water droplets, constitute one reason why the atmosphere absorbs energy at specific wavelengths. Small particles, whether transparent or not, tend to exhibit resonant characteristics, either because of their size or because of the distance between them. For example, a water droplet of a specific size will have resonant frequencies because of internal reflection (FIG. 2-5). If the wavelength of the energy is long enough so that the particles are tiny compared with the wavelength, resonant effects occur and absorption takes place at specific frequencies. Usually this is not the case at infrared frequencies; rather, the particles of dust in the air tend to scatter the energy in about the same manner for near infrared, and almost uniformly in the medium infrared range. Sometimes resonant effects can occur in the far infrared range.

Carbon dioxide and water vapor also have an absorbing effect on infrared radiation. The absorption here is because of atomic effects. Electrons will absorb energy at certain wavelengths, moving into higher-energy shells in the process. Molecules of these substances (and all others) themselves have resonant frequencies at which they vibrate if excited by energy at the right wavelength. For infrared, the atmosphere's carbon dioxide and water vapor have an attenuating effect at wavelengths of about 1.3, 1.8, 2.8, 4.3, 6.5 and 14.5 microns. There is some attenuation on either side of these wavelengths. FIG. 2-6 shows the attenuation versus wavelength in the infrared spectrum of interest. These figures are for sea level, and are given in decibels. At A, the peaks are shown as vertical lines for definition. At B, the actual attenuation graph is illustrated.

From the graph, we can see that there are certain bands in which we would not want to operate, and certain bands in which we

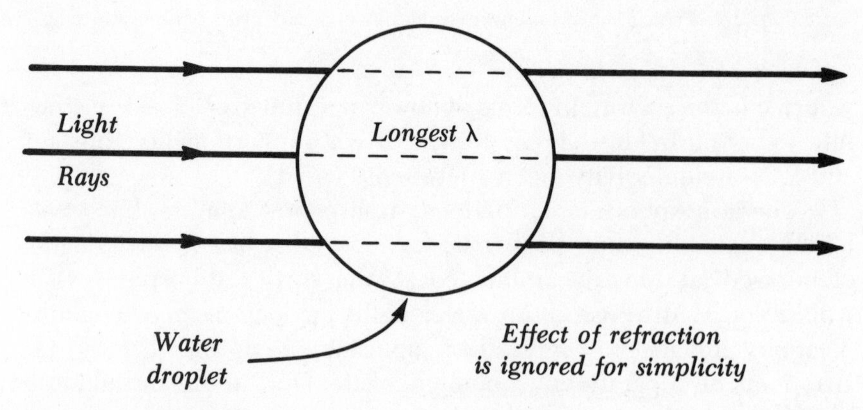

Fig. 2-5. Water droplets are resonant at a variety of wavelengths. In this simplified illustration, the effects of refraction are ignored.

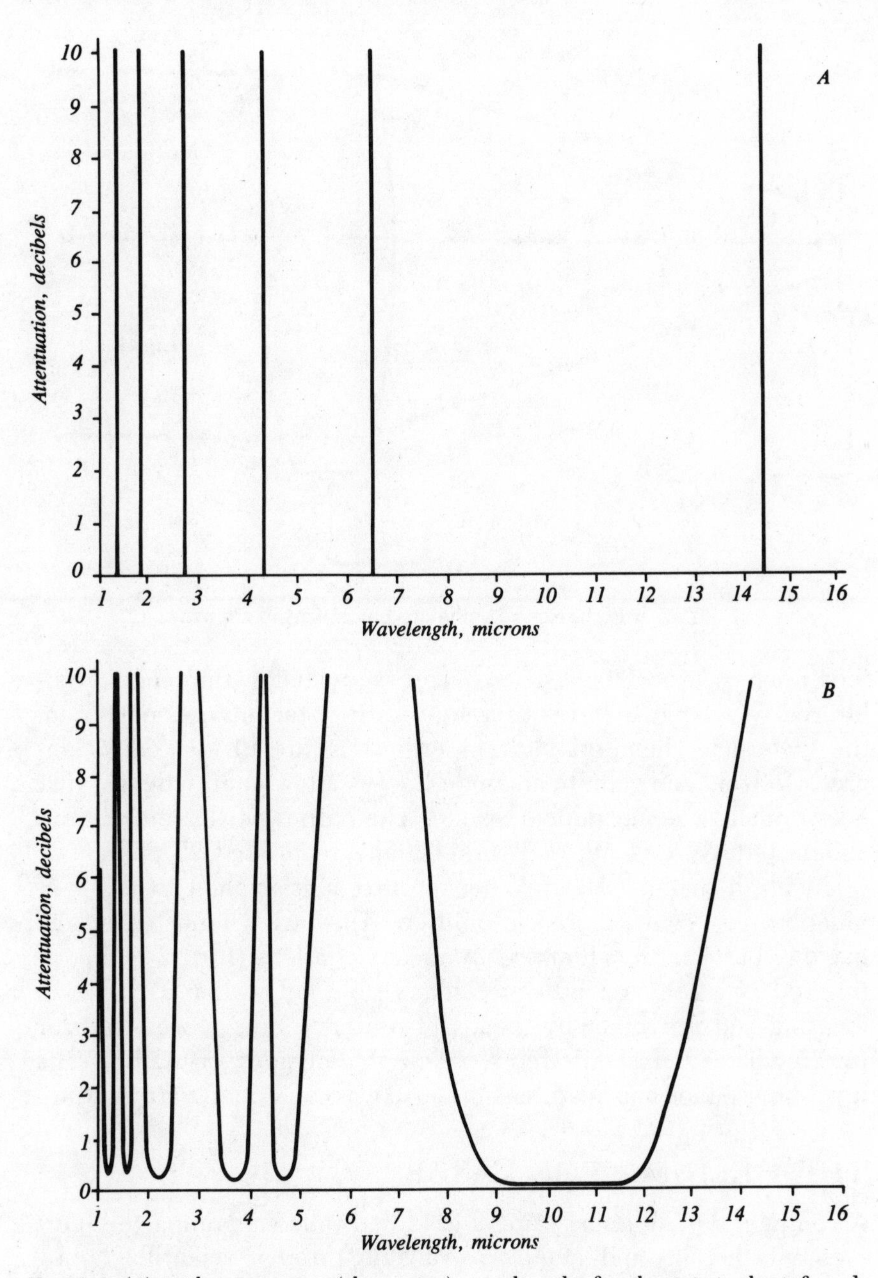

Fig. 2-6. (A) peak attenuation (absorption) wavelengths for clear air in the infrared region. (B) relative attenuation curve.

would do well. The best bands are around 1.0, 1.25, 1.7, 2.2-2.3, 3.6-3.9, 4.6-4.9, and 9.0-12.0 microns. Within these bands there is virtually no attenuation in clear air.

The gallium arsenide laser operates at a wavelength of about

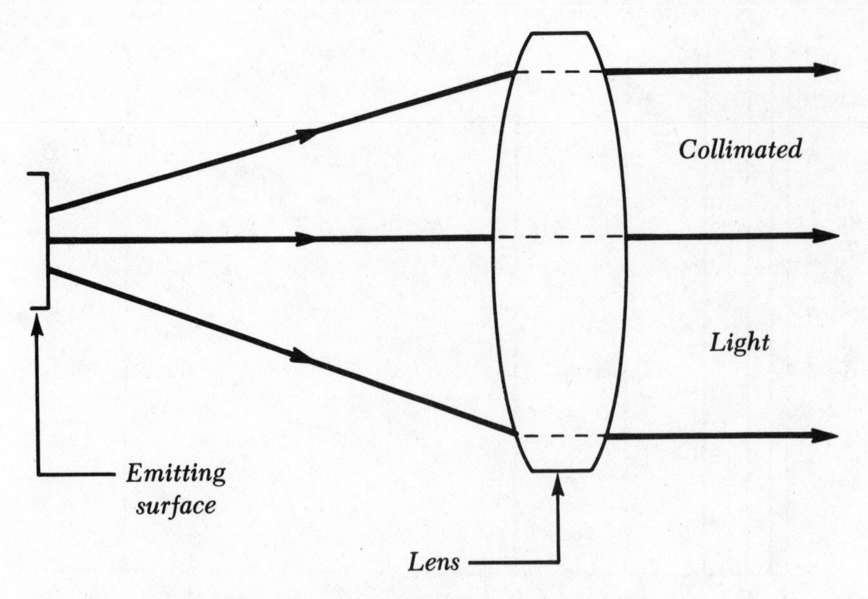

Fig. 2-7. Light can be collimated by means of a lens.

0.84 microns, or 8400 angstroms. This is just below the longest visible red wavelengths and is therefore in the near infrared portion of the spectrum. The operating temperature is a relatively cool 77 degrees Kelvin. The gallium arsenide laser is a low-energy device that is essentially a semiconductor diode. The output power ranges from about 100 mW to 1 W, and the efficiency is around 20 percent—relatively high for a laser. When a current is applied to the P-N junction, coherent infrared is emitted. The beam is not especially narrow, but can be collimated by means of a lens (FIG. 2-7). Near-infrared emissions behave much like visible light when it comes to focusing and refraction by means of ordinary glass lenses, but for intermediate- and far-infrared wavelength emissions, crystalline salts must be used for lens materials because glass and quartz are opaque.

THE HELIUM-NEON LASER

A common type of laser, using a gas-filled tube containing the inert elements helium and neon, is available through scientific hobby companies. The helium-neon laser emits red light and operates at a relatively low energy level. The gas is excited by an electrical current. Wavelengths are emitted at 0.633, 1.15 and 3.39 microns. The first is the visible red light that we see, and the other two emissions are in the infrared spectrum. The output power ranges from 10 to 100 mW—lower than the semiconductor infrared laser, and the efficiency is also lower—about five percent at the visible red wavelength.

Helium-neon laser pumping is done by means of electron collision. Energy transitions in the electron shells of the gases occur not by photon stimulation but by excitation by free electrons. The free electrons are in the form of a current caused by ionization of the gases. If a radio-frequency current is passed through the gas-filled tube, the output will be continuous. This has many advantages, especially when it comes to communications, since the continuous output is easier to modulate than a pulsed output. The helium-neon laser can be modulated by many signals at the same time, each signal having a different frequency.

Helium-neon laser output wavelength is determined by energy transitions shared by the two gases. Helium and neon have identical transitions at some wavelengths when their concentration levels are exactly at a certain value and in the correct proportion with respect to each other. The continuous-wave output has a narrow bandwidth and therefore puts more energy into a given frequency range. This is more like a true sine-wave output than many other types of lasers.

FIGURE 2-8 is a pictorial diagram of a helium-neon laser that is excited by radio-frequency energy. The tube is first evacuated, then

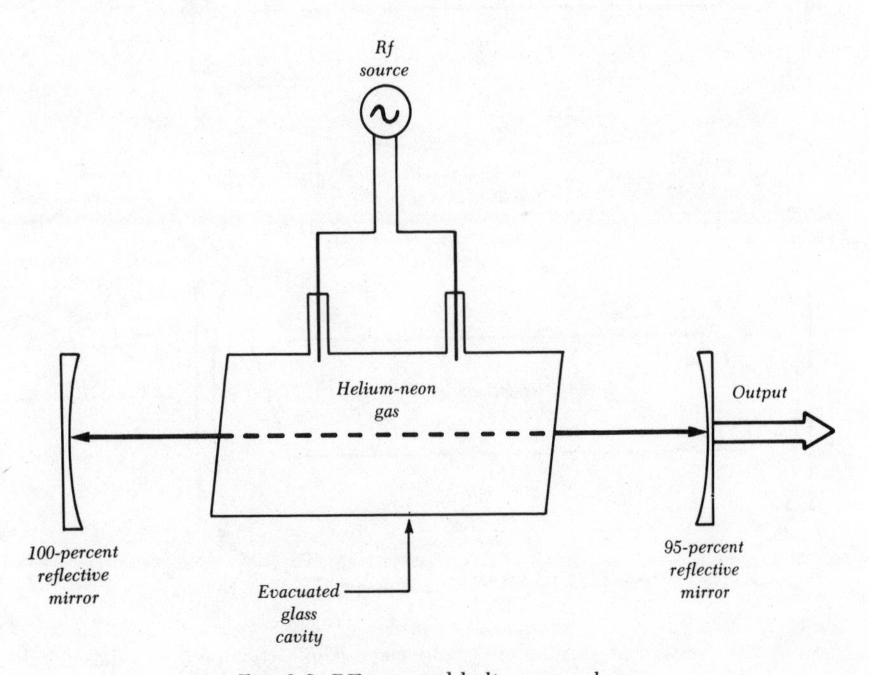

Fig. 2-8. RF-actuated helium-neon laser.

filled with helium and neon gas. The earliest helium-neon laser employed flat mirrors at each end of the resonant tube, as shown in FIG. 2-9A. More recently, concave mirrors were found to work better, placed outside the tube, the ends of which were angled as shown in FIG. 2-9B. One of the mirrors, in either case, is totally reflective, while the other is partially reflective to permit the laser beam to escape when it is sufficiently intense. The mirrors are of the first-surface type, meaning that the silvered surface is on the inside, facing the interior of the tube.

OTHER TYPES OF GAS LASERS

A mixture of nitrogen, carbon dioxide, and helium can be used to make a laser that operates at about 10.6 microns. This is in the infra-

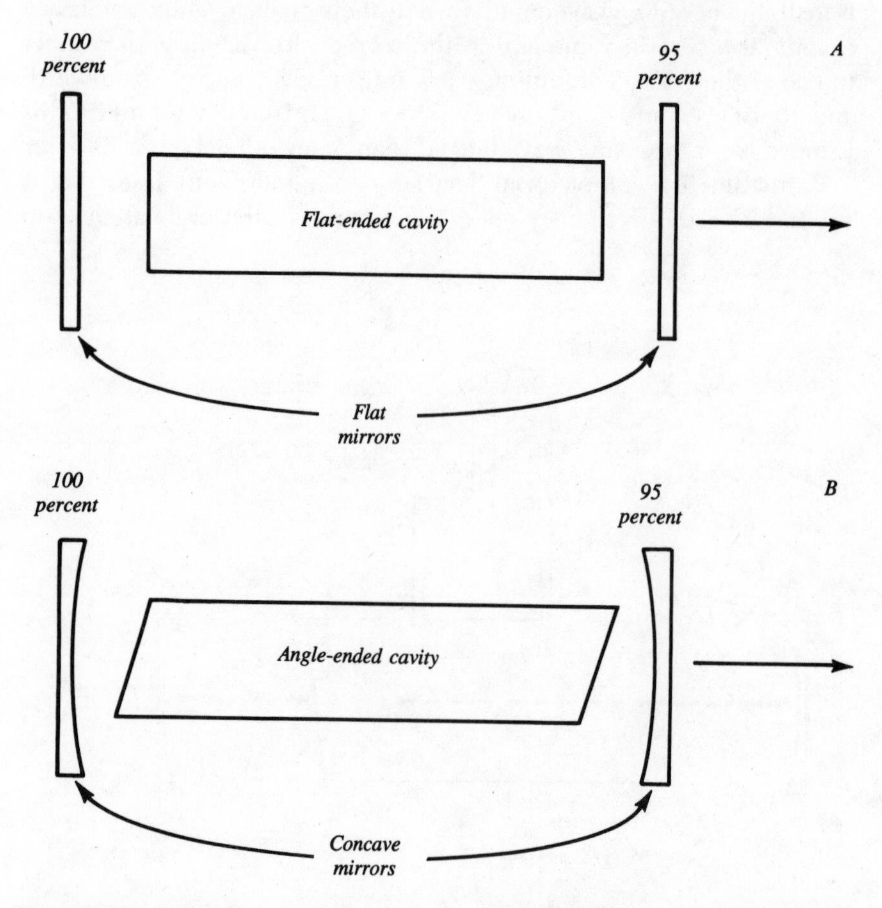

Fig. 2-9. Two types of laser cavities for gas devices. The early type used a flat-ended tube with flat mirrors (A). More recent devices use angle-ended tubes with concave reflectors (B).

red region of the electromagnetic spectrum, well below the frequency of the longest visible red light. This wavelength is about in the center of the "window" shown in FIG. 2-6B that ranges from 9.0-12.0 microns. Thus the laser beam penetrates through clear air easily. The N-CO_2-He laser can generate much more power than most lasers, with several thousand watts of continuous power possible and millions or even billions of watts in a short pulse. This laser is shown schematically in FIG. 2-10.

Many other gases, such as mercury vapor, can be used to generate laser emissions by means of electrical discharge. Hydrogen and xenon will produce ultraviolet laser energy. Some gases, such as oxygen and chlorine, can be used in lasers excited by light energy. Atomic collisions, caused by application of energy in various forms, can result in the emission of laser energy at various wavelengths in the infrared, visible, and ultraviolet parts of the spectrum. Gases expanding at high speed can also produce laser emissions. With some types of gas-expansion lasers, a continuous power output of 50 or 60 kilowatts can be obtained.

ARGON ION LASER

The argon ion laser is a gas laser operating at about 4800 angstroms, producing blue light in the visible portion of the electromagnetic spectrum. The ionized argon is kept at low pressure and the ionization, as well as the energy input, is via electrical current. The power output is generally low. The efficiency is also very low—only about 0.1 percent or less. This means that cooling is necessary even for very low-power argon ion lasers.

The pictorial representation of the argon ion laser is essentially the same as that for the helium-neon laser. The tube may have flat

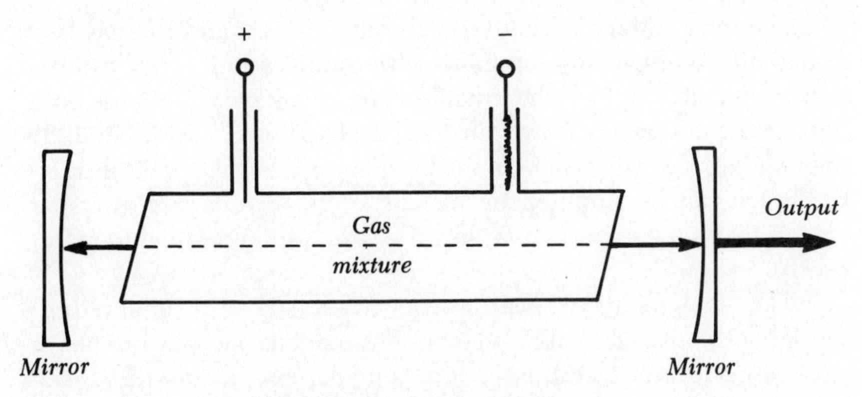

Fig. 2-10. Direct-current gas laser.

ends, and the mirrors may be flat, or the tube may have angled ends using concave mirrors for the cavity.

In high-power lasers of any type, cooling becomes an important factor. Most lasers are not very efficient. A laser that produces a power output of, say, 5,000 watts with an efficiency of five percent will require 100 kilowatts of power for excitation. This means that 95 kilowatts will be spent as heat, dissipated rather than used as energy for the laser. In a continuous-wave laser, this is a great deal of heat. The device would literally melt down if some means of cooling were not used. The whole laser can be cooled with a very low temperature liquid such as liquid helium. This will improve the efficiency as well as provide cooling for protection of the device.

LIQUID LASERS

The laser tube may be filled with an element or compound in liquid rather than gaseous state. Liquid lasers can produce energy from the near infrared to the near ultraviolet parts of the spectrum, including the whole range of visible wavelengths.

Liquid lasers are pumped by means of visible light. Sometimes the pumping source is a gas laser or solid-state laser, and sometimes it is a source of incoherent light, supplying energy in pulses. Liquid lasers may be either continuous-wave or pulsed in their outputs.

The medium in a liquid laser is usually a colored dye solution. The wavelength of the output in some liquid lasers can be varied over a fairly wide range. Such a laser is called *tunable*, and operates at a wavelength mainly determined by the color of the dye. Liquid lasers, like their gas and solid-state counterparts, must be cooled if high power output is needed, otherwise the tube and surrounding apparatus may be damaged or destroyed by heat dissipation. Typical liquid laser systems are shown in FIG. 2-11.

The main difference between the liquid laser and the gas laser is that the method of excitation is the application of an external source of visible light rather than electrical discharge, atomic collision, or expansion. As shown in FIG. 2-11A, this is usually done by surrounding the tube containing the dye solution with a flash tube. Excitation can be supplied by making both mirrors semi-transparent, so that a laser can be employed for the pumping function (FIG. 2-11B).

The neodymium-yag laser is a special crystal laser, generating a continuous wave. The abbreviation *YAG* stands for yttrium-aluminum-garnet. Actually the laser is a combination of liquid and solid materials, the neodymium being a solution inside a crystal of YAG.

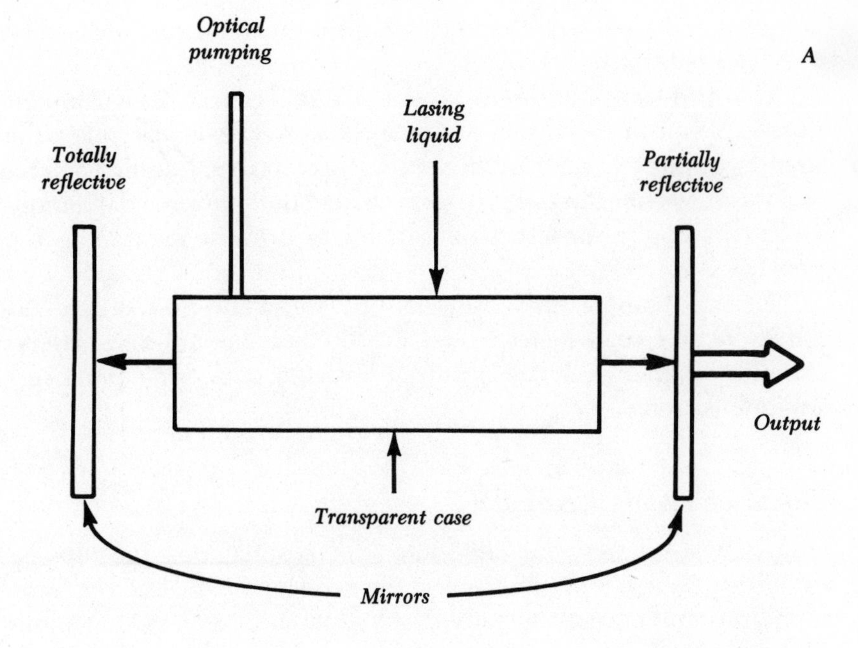

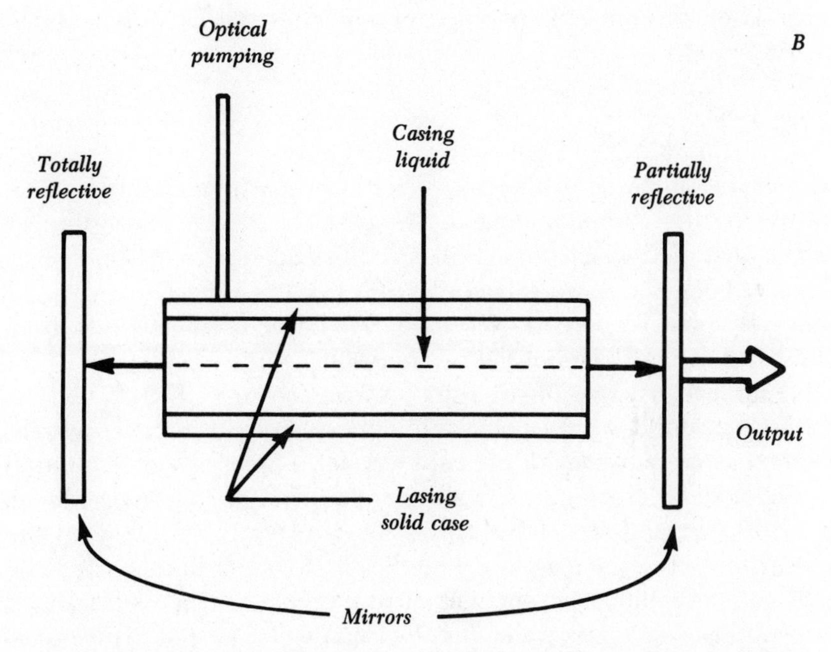

Fig. 2-11. Two kinds of liquid lasers. (A) liquid-only; (B) liquid and solid.

Energy is delivered into the laser medium by arc lamps, focused so their light falls on the crystal.

The wavelength of the neodymium-YAG laser is about 1.06 microns in the infrared portion of the spectrum. This is just below the longest visible red wavelength, so it is in the near infrared. The efficiency is low—on the order of 1 percent. High-power neodymium-yag lasers require a method of cooling to prevent damage to the crystal.

The neodymium-YAG laser can operate in the pulsed mode as well as in the continuous mode. Pulsed neodymium-YAG lasers, with much higher peak power output levels, are used in certain surgical applications.

SOLID-STATE LASERS

Solid-state lasers can be divided into several categories: the ruby laser, which is an early type of laser, the semiconductor laser (gallium arsenide is one example already discussed), and lasers using various solid dielectric materials such as plastics and crystalline salts. Solid-state lasers are optically pumped in a manner similar to that shown in FIG. 2-11A.

The Ruby Laser

The "ruby" material in the ruby laser is primarily aluminum oxide, with a trace of chromium added. The flash tube is a cylindrical piece of this material, which appears as a reddish solid with reflective surfaces at each end. One reflector is 100 percent reflective while the other is about 95 percent reflective. The laser beam emerges from the partly silvered end of the ruby crystal.

The construction of the ruby laser is shown in FIG. 2-12. The flash tube supplies pulses of visible light that cause the energy transitions in the electrons of the ruby crystal. Some energy is emitted in the form of laser light, at a wavelength of 6943 angstroms, which is at the red end of the spectrum. The energy is pulsed, with the power output ranging from one milliwatt to 10 milliwatts, and the efficiency is rather low—only about 0.2 percent. This means that a one watt power input yields just two milliwatts output. The rest is dissipated as heat or wasted as escaped light from the flash tube.

There are hundreds of thousands of standing waves within a typical ruby laser. Thus the cavity, or solid, has many resonant frequencies that are very close to the wavelength of 6943 angstroms. The output of the laser is in the form of many *spikes*, with one group

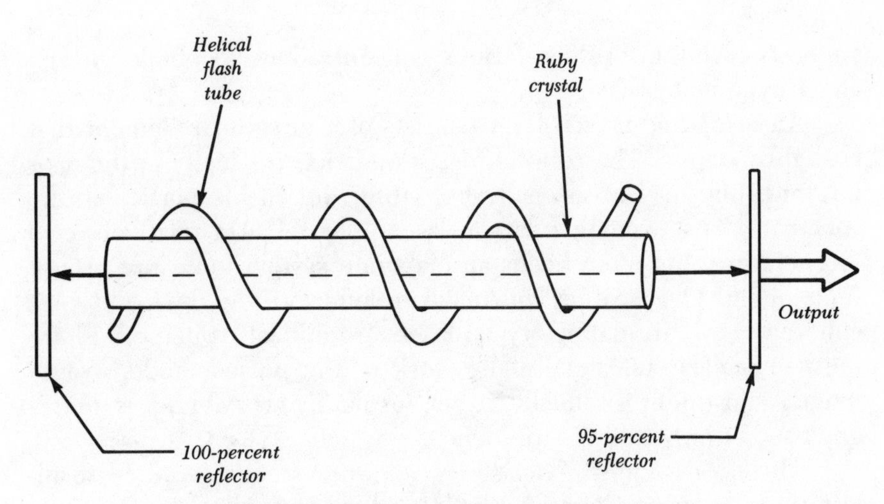

Fig. 2-12. Simplified pictorial diagram of a ruby laser.

of *spikes* for each pulse of the pumping flash lamp. The individual *spikes* are about one or two microseconds long. The entire laser pulse lasts about four to six tenths of a millisecond (4×10^{-4} to 6×10^{-4} second).

More advanced ruby lasers can be operated continuously rather than in the pulsed mode. Sometimes a second ruby laser is excited by the first one, the second laser acting as an amplifier to obtain greater power output. This is done in the pulsed mode and a peak power output of millions of watts can be realized. (The average power is of course much lower than the peak power in a pulsed laser.) The pulses last for only a few nanoseconds.

The Insulating Crystal Laser

Different elements can be employed as the cavity material in a solid laser. A common material is a mixture of gallium arsenide and gallium antimonide. Mixed crystal lasers have a special property: their wavelengths can be adjusted. This is accomplished by varying the relative concentrations of the different substances in the cavity medium. In the case of the GaAsSb (gallium-arnsenide-antimonide) laser, the wavelength can be tuned continuously from about 9,000 to 12,000 angstroms in the near infrared end of the spectrum.

Other common materials used in mixed crystal lasers include gallium–arsenide–phosphide, aluminum–gallium–arsenide, and pure gallium–arsenide. Semiconductor crystals are grown in a liquid solution in a manner similar to ordinary semiconductor diode and transistor crystals. The semiconductors are *doped* (mixed with trace amounts of impurities) in various concentrations to obtain N-type and P-type materials.

An insulating crystal laser consists of a rare-earth element that is slightly doped. The resulting laser emission is usually in the near infrared, buy may occasionally be within the visible portion of the spectrum. The insulating crystal laser operates at a relatively cool temperature, but often needs an elaborate system to ensure proper functioning. Overheating will result in failure of the device to emit coherent light. Insulating crystal lasers are optically pumped, as are most crystal lasers, and usually work in the pulsed mode. Wavelength adjustment is difficult, so the insulating crystal laser is generally used at only a single frequency.

The neodymium-YAG laser is sometimes considered an insulating crystal laser. Glass may be doped with neodymium to obtain laser operation with sufficiently intense optical pumping. With proper temperature (20 degrees Celsius or less) and intense pumping, the efficiency of such a laser is about 3 percent with materials of high optical quality. The peak output power may be on the order of 10^{12} (one trillion) watts or even, under optimum circumstances, 10^{13} (10 trillion) watts. The actual duration of the pulses is very short, making the average power just a few watts. This is readily apparent simply from the mention of such high peak power—a continuous trillion-watt generator could supply 100 million average households simultaneously, and this is about all the power consumed at any one time by the entire civilized world.

SEMICONDUCTOR LASERS

The semiconductor laser is a special form of solid state laser. Many semiconductor substances act as light-emitting sources when they are forward-biased. The familiar light-emitting diode is a good example. Some light-emitting diodes produce coherent light and therefore can aptly be called laser diodes. Gallium arsenide, introduced above, forms a laser when doped in such a manner as to create a P-N junction. The GaAs laser diode is probably the most common example of a laser diode. GaAs laser diodes are available in electronics stores and thus are well suited to home experimentation.

The GaAs laser diode uses electrical energy to produce coherent light directly. The output is in the infrared part of the spectrum. FIGURE 2-13 is a simplified pictorial drawing of a typical semiconductor laser. The clear layer makes it possible for the infrared light to escape.

If modulated direct current is applied to a semiconductor laser,

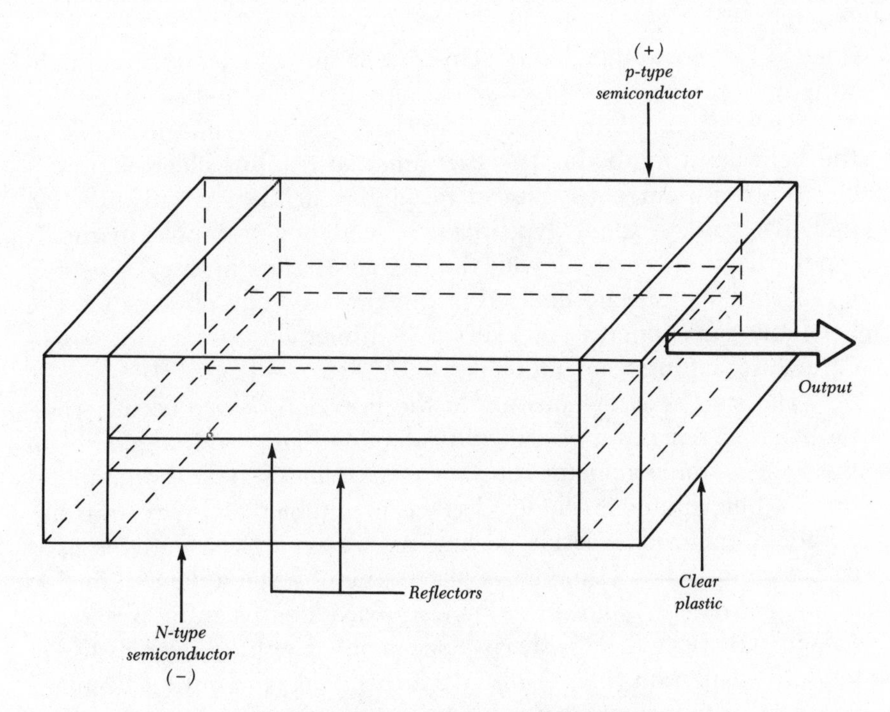

Fig. 2-13. Semiconductor injection laser.

a modulated-light communications transmitter can be made. The coherent light has the advantage that it can be easily focused into a nearly parallel beam. Using large diameter reflecting mirrors at both the transmitter and receiver, communications can be realized over long distances through the atmosphere or empty space. The laser has the advantage that it can be modulated to almost 100 percent, as compared with incansescent or fluorescent lights of the noncoherent type which can be modulated to only a few percent. This characteristic in effect makes the laser more efficient for communications purposes, as compared with conventional light sources.

The smallest and most modern type of semiconductor laser is called the injection laser. The injection laser is similar to a solid-state crystal laser, but has the advantage of being able to use electrical energy directly. A crystal laser generally needs to be optically pumped. The gas laser raises the kinetic energy of electrons in atoms, and this in turn produces laser light. The directness with which the injection laser works results in a generally higher degree of efficiency than is possible with most other lasers. The efficiency may be as high as 40 percent, compared with 1 percent for an optically pumped laser and 0.01 to 0.1 percent for a gas laser excited by means of electrical currents.

The equipment used with injection lasers is simpler than that needed for most other lasers. This makes for compactness and light weight.

The injection laser operates on basically the same principle as the light-emitting diode. The P-N junction exhibits electroluminescence when a current is passed through it in the forward direction (positive pole to the P type material and negative pole to the N type). Electrons combine with the charge carriers in the P type material, and photons are emitted in this process. The photons tend to have the same energy for every recombination, producing monochromatic, infrared, or ultraviolet light (FIG. 2-14).

The start of laser emission in the injection device occurs when the current reaches a certain critical value. This value depends on the type of semiconductor substance used, and also on the temperature. When gallium arsenide (GaAs) is maintained at a very low temperature, infrared emission occurs. An observing apparatus is used to determine when lasing takes place. The observing device consists of a spectrometer, allowing a band of wavelengths to be observed. Lasing is indicated by a sharp peak in the amplitude-versus-wavelength distribution (FIG. 2-15). With this peak, the output becomes practically monochromatic and coherent.

Compared with other types of lasers, the injection laser has certain disadvantages. The power output is limited since the junction must be small. The output tends to have a broader bandwidth than, for example, a ruby or helium-neon laser. These disadvantages offset the advantages of low power consumption, easy availability, and ease of modulation. The diode laser emits a broad beam, but this can be focused using a glass lens. The GaAs injection laser produces infrared in the near region, close enough to the visible spectrum so

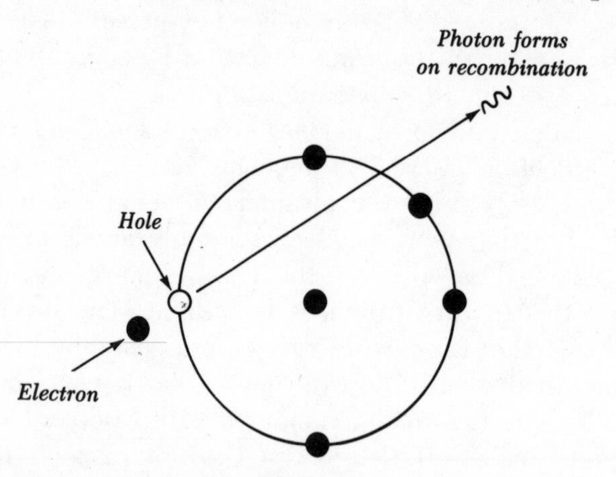

Fig. 2-14. Emission of a photon may occur when an electron fills a hole in a semiconductor material. This is called recombination emission.

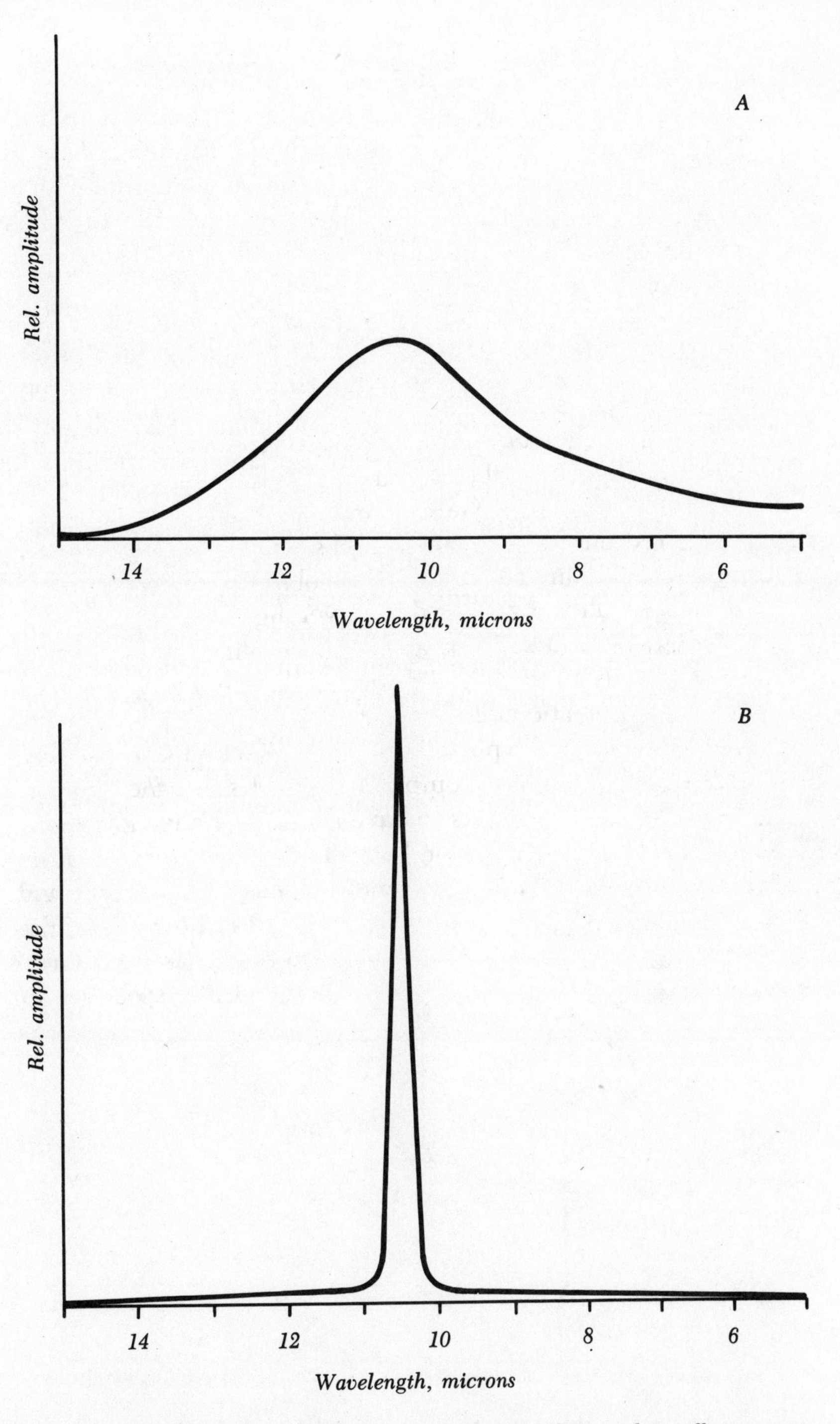

Fig. 2-15. Infrared emission of gallium-arsenide injection laser with insufficient excitation (A) and sufficient excitation (B). Lasing is characterized by a sharp peak in amplitude with narrow bandwidth.

that ordinary lenses can be used. (The focal length of a glass lens is a little different for near infrared than for visible light).

A simple circuit for home experimentation is shown in FIG. 2-16. The transistor may be any type capable of handling about 10 amperes. This circuit produces pulses that can be monitored with an oscilloscope. The pulses should be about two times the threshold current of the laser diode. The output can be observed using an infrared detector or viewing card.

Gallium arsenide diode lasers are available from RCA in various amperage levels. The circuit just described is suitable for use with the S62001 type, with a peak of about 10 amperes. The S62003, S62006 and S62009 types have peak current ratings of 25, 40 and 75 amperes respectively. It is necessary to use larger transistors with these diode lasers to accommodate the larger peak currents.

Ideally, focusing is accomplished by means of a parabolic reflector of the first-surface type, similar to that found in a reflecting telescope except with a shorter focal length. The reflectors used with large spot lights are ideal, but expensive and hard to come by. Fresnel type lenses are available from Edmund Scientific in diameters up to 1 foot and may also be used. By using an infrared detector and demodulation system, and modulating the laser itself via audio in the circuit shown here, an infrared communications system can be made. We will look more closely at communications systems in another chapter.

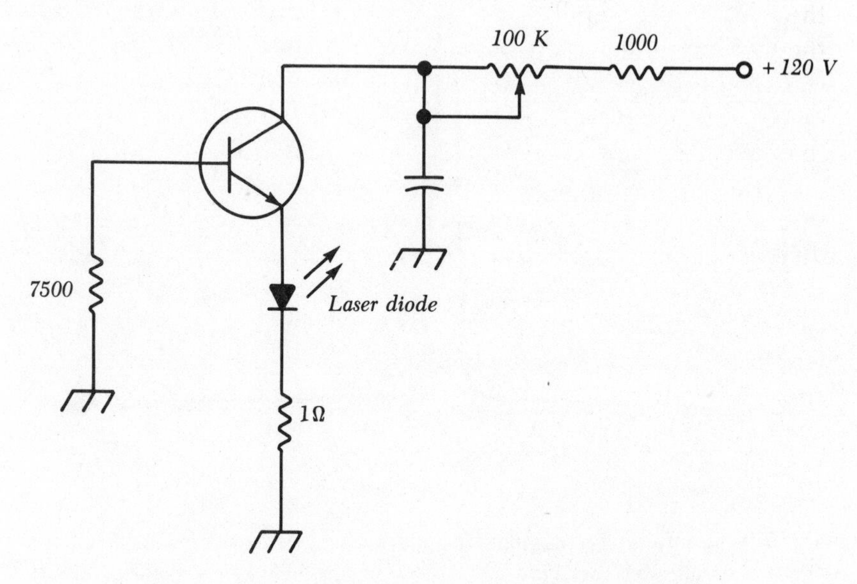

Fig. 2-16. Simple pulse circuit for GaAs injection laser.

Various optical laser diodes are available from consumer sources. The circuits used for them are identical with the one shown here. Some diodes can be operated from a continuous current source instead of a pulsed source such as that shown here.

DIFFERENCES IN LASERS OF THE SAME CATEGORY

Gas lasers, liquid lasers and solid-state lasers have decidely different lasing media and operate according to different principles, but even within the same type classification, two lasers may be different in terms of power output, beamwidth, wavelength and energy distribution.

Power output depends on the physical size of the laser, both in terms of length, width, and also on the intensity of the pumping source. In some lasers the power output is limited by more practical constraints, such as the limitations of crystal growing techniques. The neodymium-YAG laser is an example of this—the output cannot exceed about 200 watts. Power output is measured in watts-per-meter of discharge for lasers excited by electrical currents. A CO_2 gas laser has about 50 to 90 watts of output per meter of discharge. There are many sizes of lasers that will produce an output of a given wavelength, since usually the laser medium is many thousands of times the length of a wavelength.

The beamwidth of a laser depends on the physical diameter of the medium. Generally, smaller media produce narrower beams, although larger diameter lasing cavities produce beams that spread less rapidly as distance from the laser increases. Thus at a great distance, a larger diameter laser may actually have a narrower beam than a small-diameter laser of the same type (FIG. 2-17).

The wavelength of a laser depends on the resonant frequency of the cavity as well as the natural resonant frequency of the lasing medium. Some media have more than one natural resonant frequency, so the one at which lasing takes place depends on the length of the cavity. Some lasers are continuously tunable over a range of wavelengths. Most lasers operate in the infrared or visible part of the spectrum. Some operate in the ultraviolet, and research is still proceeding in the X-ray band.

The energy distribution within a laser beam is determined according to a cross-sectional method, and can be largely classified as either the "spot" or "donut" type described earlier. In either of these two primary modes, the relative concentration may vary considerably. For example, in the spot mode, the energy may fade off

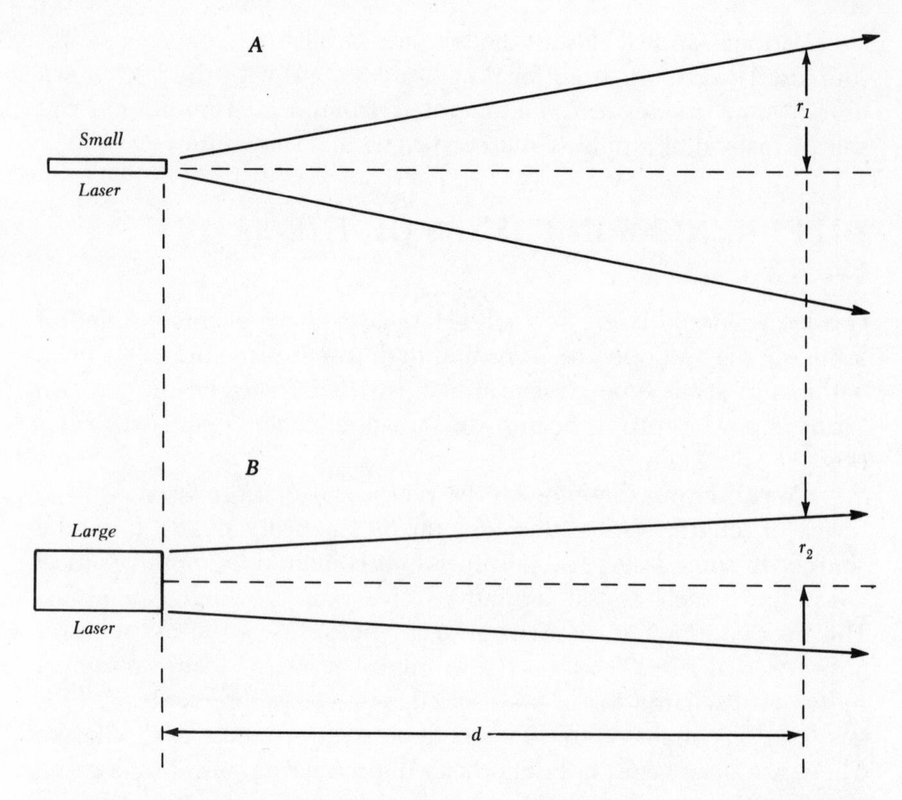

Fig. 2-17. A small laser (A) may produce a wider beam at a great distance, d, than a larger laser (B). The small laser produces a beam of radius r_1 and the large laser a beam of radius r_2.

gradually and evenly with increasing distance from the center, as shown in the diagram of FIG. 2-18A, or it may drop off abruptly as shown at B. The same differences apply to the donut mode. The energy distribution may appear as an almost perfect ring, with most of the energy between two concentric circles, or it may be more diffused. In either the spot or donut mode, the energy distribution tends to become more and more diffused as the distance from the source is increased. This is because of natural geometric scattering, and also because of atmospheric scattering or diffusion under water or whatever other medium the laser passes through.

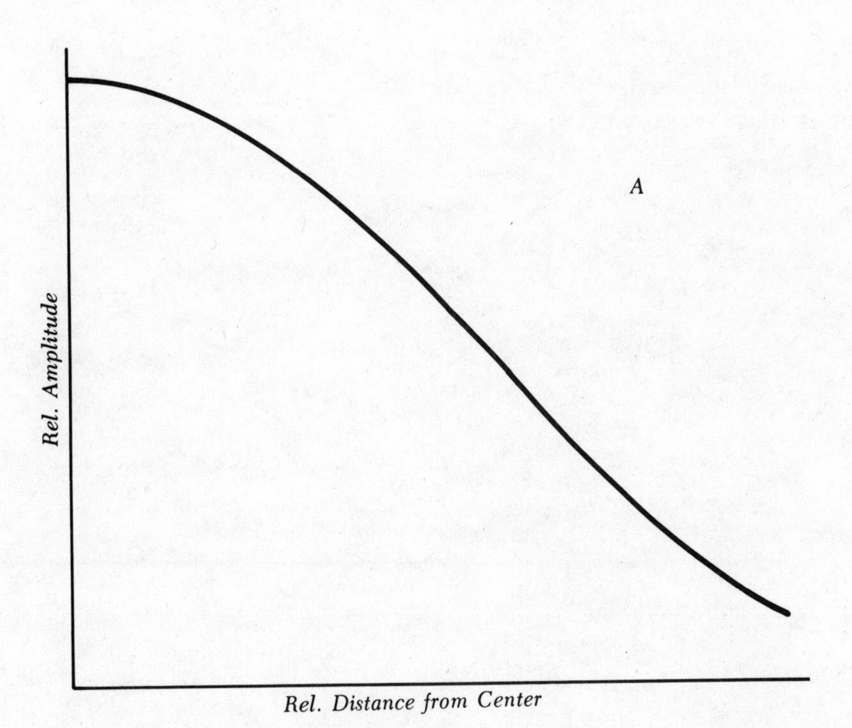

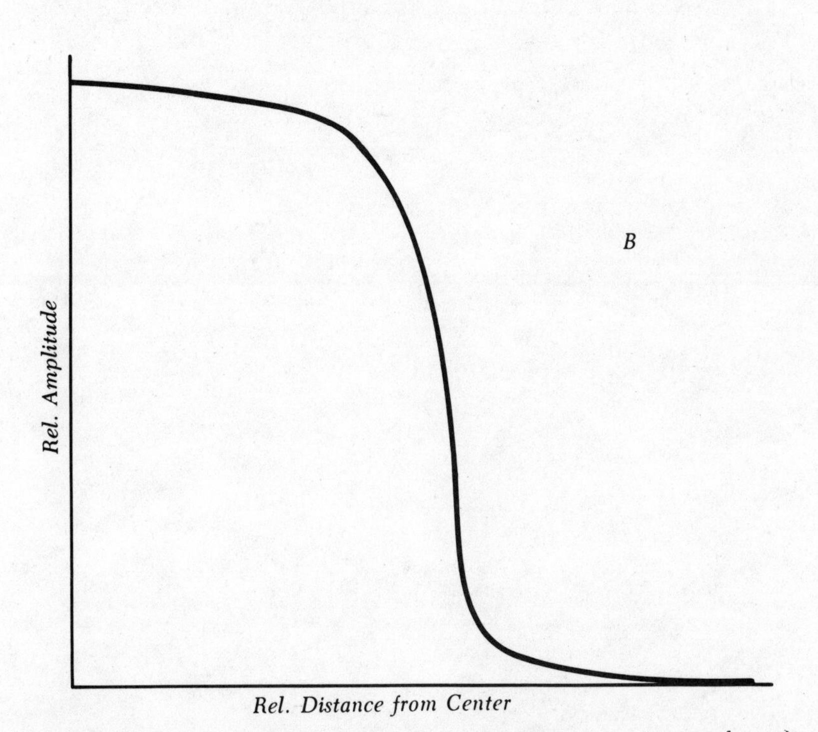

Fig. 2-18. Distributions in a "spot" laser may be diffuse (A) or have a sharp dropott (B) with increasing radius from the beam center.

3

Lasers in Medicine

When we think of lasers, images arise of "photon torpedoes," capable of destroying massive star ships, or the continuous blast of the "phaser," set to kill and backed up with megawatt-hour power supplies of lithium crystals. How ironic that there are such things as "phasors" and "lithium batteries" today, although neither are of the horrific magnitude portrayed in *Star Trek*. This is not to discount such possibilities altogether, though, as we will see.

Lasers exist today that can burn through solid metal and which can be used to kill and destroy. The plan for such devices began as soon as the laser's ability to concentrate power was theoretically shown and practically demonstrated. Any new technology is bound to have its following of those interested in its use to defend their interests or lives. Lasers are already being used to make life better, if not to forcibly keep us free. This significance has been apparent for some time now in the field of surgical medicine. More recently, diagnostic medicine has entered into the scenario as well.

The idea for using a laser as a surgical cutting tool is straightforward enough. By simply concentrating the power, tissue can be "burned" apart. The practical design of surgical lasers involves concentrating the beam into a fine cross-section so that only the appropriate areas are cut.

Lasers can also be used for diagnostic purposes. Coherent light makes it possible to view large molecules and even viruses in action. Molecules, viruses, bacteria, or cells can be traced, and in some cases their motion can be controlled.

It is possible that someday lasers may be employed to control genetic behavior. This could make it possible, for example, to correct potential birth defects. It is furthermore possible that some mutations could be deliberately engineered, such as a mental predisposition towards the monotonous lifestyle of a foot soldier or even the peculiar genius of those whose life's work is but to manufacture destruction.

LASER EYE SURGERY

Once of the earliest surgical applications of the laser was in the repair of a detached retina. Other eye problems can also be corrected by means of fine lasers. The effect of a surgical laser is determined by the temperature to which the laser beam heats the tissues, and also on the amount of time that the tissue is exposed to the light.

The normal body temperature is 37 degrees Celsius. In a process called photoradiation, the temperature of a given sample is raised to 38 degrees Celsius. The exposure time is about 30 minutes. This process is used to destroy eye tumors. In photocoagulation, the temperature is raised to about 65 degrees Celsius. Photocoagulation causes bleeding to stop, such as in a retinal hemorrhage. A high-energy laser can raise the temperature to hundreds or thousands of degrees Celsius, and these processes are known as photovaporization (about 400 degrees Celsius) and photodisruption (about 20,000 degrees Celsius). The process of photovaporization can be used to destroy malignant tumors, removing all the cancer cells so none remain to spread. Photodistruption causes tissue to part, and the action is like that of a tiny scalpel. Specifically designated cuts can be made in this way for removing tumors or for other surgical procedures on a microscopic scale. These high-energy lasers are used for periods measured in microseconds (millionths of a second) or nanoseconds (billionths of a second).

Besides the removal of tumors and the repairing of broken blood vessels or detached retinas, laser eye surgery is used to place artificial lenses to improve vision. This is done in cases of cataracts, where lenses have become "milky," thus interfering with sight. For relief of glaucoma, or excessive pressure in the eyeball, a tiny hole is burned in the iris, the colored part of the eye around the black pupil, allowing excess fluid to escape.

The main advantages of laser eye surgery over more conventional methods are increased precision (a given area can be vaporized or cut without affecting any other area), painlessness, obviating the need for anesthetics, and reduced risk of accident.

THE ENDOSCOPE

Laser beams can be used for diagnostic purposes as well as surgical operations. The interior of the body can be directly viewed by means of a device called an endoscope, thus making exploratory surgery unnecessary. FIGURE 3-1 is a drawing of an endoscope.

The device consists of a flexible tube containing an optical fiber that carries a beam of light illuminating the region in the vicinity of the end of the tube. Other openings provide for dispersal of gas and removal of fluid. There is a forceps at the end of the tube, controllable for the removal of tissue for later biopsy or other analysis. A laser, guided via another optical fiber, is used for surgery or removal of tissue for biopsy. The laser can be directed by viewing through a return optical fiber, so that its energy is precisely directed. The laser advice may cost up to $150,000, which, in part, explains why rising medical costs are such an issue today: the sophisticated apparatus now available is itself very expensive.

The endoscope is useful in repairing ulcers in the stomach. The device is inserted into the mouth or nose, and run down the esophagus. Bleeding ulcers or burst blood vessels in the esophagus can be repaired. Tumors can also be repaired. It may be possible to save lives using this technique in the case of a hemorrhage in one of the

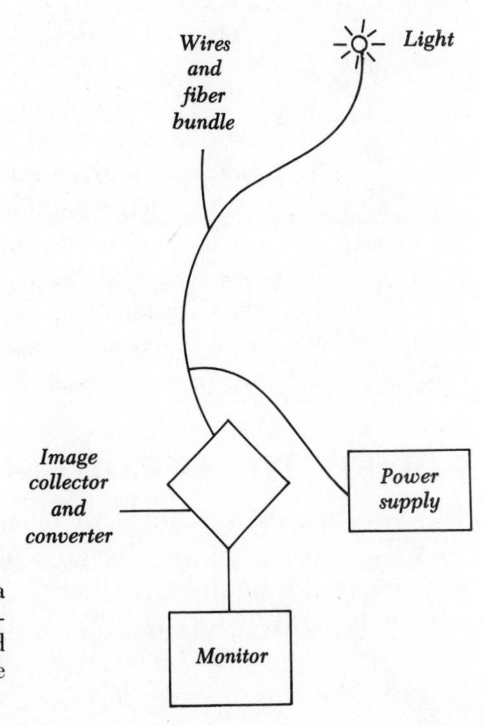

Fig. 3-1. Simplified diagram of a endoscope. The fiber bundle carries the light for illumination, and also carries the image back to the viewing monitor.

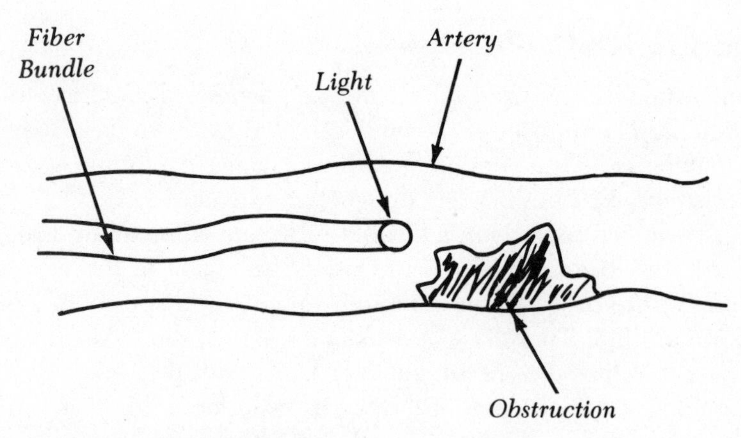

Fig. 3-2. Removal of an arterial obstruction using a laser is simple if the artery is straight and fairly large in diameter.

varices that often form along the esophagus lining in chronic alcoholics. In the past, such crises have often been fatal, because internal bleeding does not stop readily.

Damaged lungs can also be examined using the endoscope. The tube is inserted into the bronchial tube. Fluid can be removed in cases of pneumonia, and small samples of lung tissues can be removed for biopsy. This allows both diagnosis and treatment of pneumonia and other lung disorders.

In the large intestine, endoscopy allows the diagnosis of such conditions as diverticulosis, diverticulitis, cancer, acute colitis, and other conditions. The patient is given a series of enemas followed by insertion of the endoscope into the rectum. This technique eliminates the use of radiation in diagnosing intestinal problems.

The endoscope has made possible surgery that was inoperable previously. Certain kinds of brain tumors and spinal tumors can be removed without causing other permanent damage. It is possible to visually explore the interior of an artery or vein, finding and correcting conditions in which obstructions may impede circulation or in which blood clots present a risk (FIG. 3-2).

HOLOGRAPHY AS A DIAGNOSTIC AID

The laser makes it possible to obtain a three-dimensional view, known as a hologram, of a scene or object. The technical details of holography will be discussed later. Right now, the important concept is the realistic reproduction, in three dimensions, of an object or group of objects, and the way in which the view of the scene can be varied to correspond to different points of view. Medical doctors have long been seeking ways to obtain three-dimensional reproductions of the interior of the human body, without actually having to cut the body open. This has been done by means of X rays, providing series of cross-sectional views of organs such as the liver. This is also now being accomplished by a technique known as nuclear-magnetic resonance imaging (NMR or NMRI).

Laser holography has also been employed to determine the proper time for removing sutures (stitches) from the eye. In a cornea transplant, the sutures must be removed at just the right time. If the sutures are taken out too soon, astigmatism—distortion of the lens and focusing point on the retina—may occur. If the removal is put off too long, it becomes more difficult, and there is an increased risk of infection. The ideal "window" is a rather short time and can be discovered by means of holography. A side view of the cornea allows determination of its strength without any physical contact. A helium-neon laser is used. The beam is split by mirrors, one partially reflecting, into two parallel beams. The two beams are then used to generate the hologram. The resulting three-dimensional picture is compared visually with the ideal shape of the cornea. Any deviation from this ideal is easily seen and indicates that the cornea has not yet completely healed. The test is performed at frequent intervals until the healing is complete as indicated by the hologram. The sutures can then be removed without risk of astigmatism. The holography technique can be used to determine the exact nature of astigmatism in the eye.

Holography can also be employed as an early means of detecting possible breast cancer. Basically this consists of generating an interference pattern that readily shows any irregularity of the skin surface, such as a bump, that might pass unnoticed in ordinary light and be too small or too soft to be felt by conventional examination. An irregularity does not necessarily indicate that there is a cancerous tumor, but does show that the circulation is nonuniform in a given area and this could indicate a possible tumor.

It is possible that short-wavelength holography may be employed someday to obtain, for example, a three-dimensional X-ray view of certain internal parts of the body.

LASER SPECTROMETRY

The human body receives and processes chemicals from air, water, and food throughout a lifetime. The chemical processes can be evaluated by various means to determine whether or not the body is working properly. Some substances are normally excreted in the breath, urine and feces in certain amounts and a significant change may indicate trouble. Radioactive isotopes of various chemicals have been used in medicine as "tracers," being easy to follow through the body. But radiation exposure has been of increasing concern recently, and many times it is difficult or impossible to get a radioisotope that will function well in this role.

Non-radioactive isotopes can be used as tracers, but until recently it has been difficult to detect such chemicals. Mass spectrometry has been used, but the process is quite complicated, involving careful preparation of samples and very expensive equipment. A technique has been developed for spectrometry based on the degree of infrared absorption resulting from the presence of certain chemicals. Drs. Peter Lee and Richard Majkowski have applied a tunable infrared diode laser to this task. The device was originally developed by General Motors for the purpose of determining the relative amounts of various gases in automobile exhaust. The use of this tunable diode laser has resulted in excellent sensitivity and resolution as a spectrometer.

The laser spectrometer works according to the same principle as the laser used by astronomers to determine the composition of interstellar gases—by means of the absorption spectrum. A particular gas has characteristic wavelengths at which it attenuates electromagnetic radiation much more than at adjacent wavelengths. This happens because, at certain frequencies, photons striking the atoms cause an electron to move to a higher-energy orbital shell (FIG. 3-3). When there are several isotopes of a certain gas present together, the absorption wavelengths are different for each isotope, but only by a very small amount. Using conventional spectrometry, the difference is difficult or impossible to see. But with the tunable diode laser, which emits energy at a discrete, precisely known, and controllable wavelength, this presents no problem. The spectral resolution is several hundreds or thousands of times smaller than the spectral line spacings.

The system employed by General Motors used a single crystal with layers of doped lead telluride and an alloy of lead–europium–selenium telluride. Infrared radiation was collimated, passed through the sample of carbon monoxide gas, and focused onto an infrared detector. Wavelength modulation and harmonic detection enchanced the sensitivity of the system. Tuning the laser produced a curve such as that of FIG. 3-4, which is a portion of the spectrum for exhaled human breath.

According to Dr. Lee, the infrared spectrometer may have many diagnostic uses in medicine. The most obvious is determining

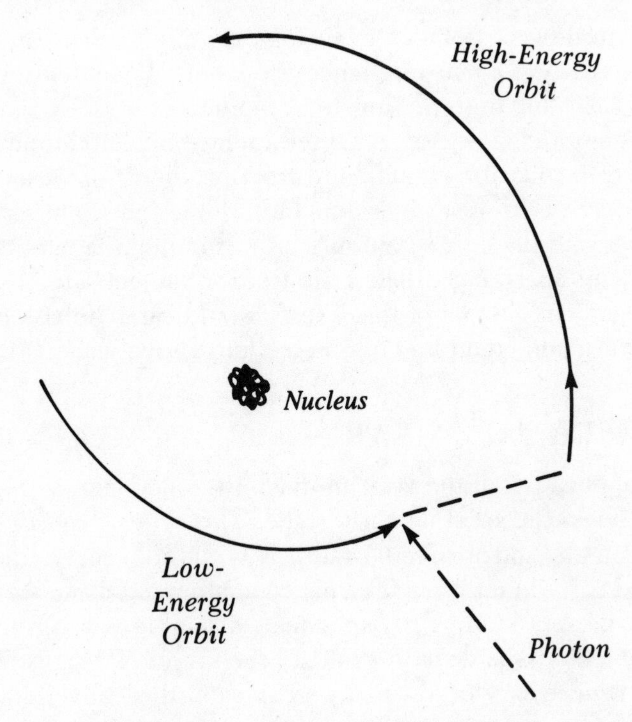

Fig. 3-3. When a photon strikes an electron in an atom, the electron gains energy.

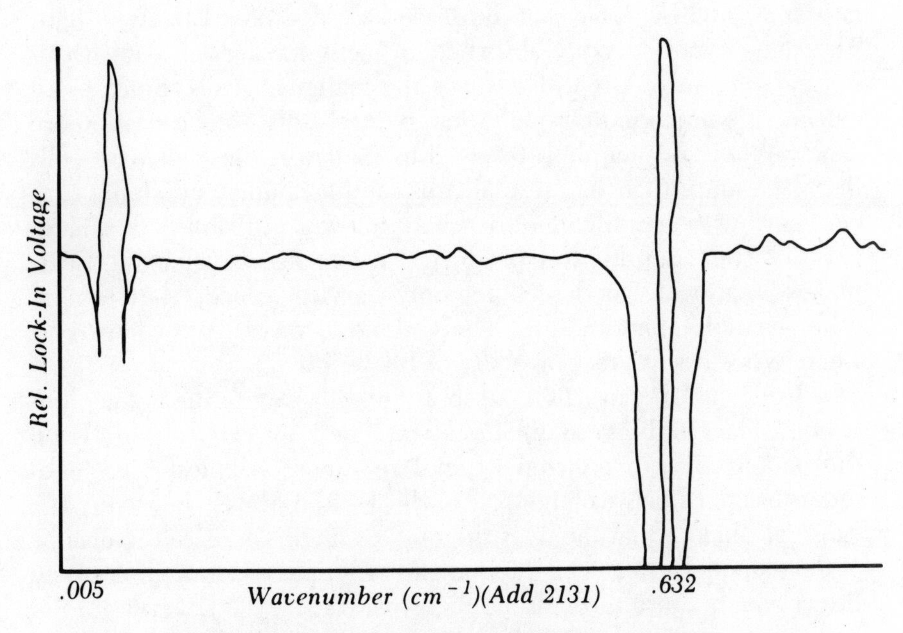

Fig. 3-4. Laser spectrograph of exhaled human breath.

the concentration of isotopes in biological samples that can be converted into carbon monoxide, since the General Motors device was used primarily for determining the amounts of various isotopes of this gas in automobile exhaust. Other applications might include the diagnosis of fat malabsorption, intestinal problems of various sorts, alcoholic liver cirrhosis, liver function, lung function, nutritional assesments, and diabetic conditions. For example, diabetes could be diagnosed by analyzing exhaled air from a subject administered a tagged sugar sample. All of these tests would carry no risk of radiation exposure and could be done in a comparatively short time.

THE LASER PAP SMEAR

Another application of the laser in diagnostic medicine is the detection of cervical cancer at an early stage. This type of cancer is common in women, and it is important, as with any cancer, that it be discovered early to improve the chance of stopping its spread and to reduce surgical risks. In the pap smear, a sample is taken from the cervix and a biopsy is done to evaluate the sample for signs of malignancy. It is necessary to test many samples in this conventional procedure: there is sometimes difficulty telling whether a given sample indicates cancer or whether it is a representative of a severe but nonmalignant inflammation.

A technique using a laser provides an improvement. A sample is first stained by a dye that fluoresces when exposed to laser light. The dye is more readily absorbed by cells having a greater-than-normal amount of DNA. These are the malignant cells. Since these cells divide much more rapidly than normal cells, they contain more DNA. When the sample is exposed to the laser, the malignant cells fluoresce more than the normal cells, and a sample containing cancer can thus be readily identified with a lower probability of a "false alarm." The laser light is produced at various wavelengths. This makes it possible to detect not only cervical cancer, but several other possible abnormalities. The technique was first developed by Leon Wheeless at the University of Rochester.

In China, the laser is used in a different way to distinguish between cancer and severe inflammation. The laser is used directly on the patient. Dye is injected without removing samples. This technique has not been widely used in the United States, however, because of concern that the dyes themselves might increase the chance of developing cancer. The dye, taken orally, passes through not only the cervix, but also through the digestive tract and other parts of the body.

TREATMENT OF CANCER

Lasers can be used as a scalpel, to cut out tumors or to remove obstructions from arteries and veins with minimal intrusion. However, lasers can also be employed to treat cancer without actually cutting it out. This technique is similar to that used to detect cancer. Dyes which are absorbed more readily by cancer cells than by nonmalignant cells are injected into the body. The tissue is then subjected to the light of a laser, at a wavelength that is absorbed by the dye. This may or may not be a wavelength within the visible spectrum. The dye is heated by this absorption to a temperature sufficiently high to kill the cancer cells. The normal cells, containing less dye, are heated much less and are not affected (FIG. 3-5). Of course it is necessary to ensure that the intensity of the irradiating laser beam is not high enough to damage normal cells, while still being enough to

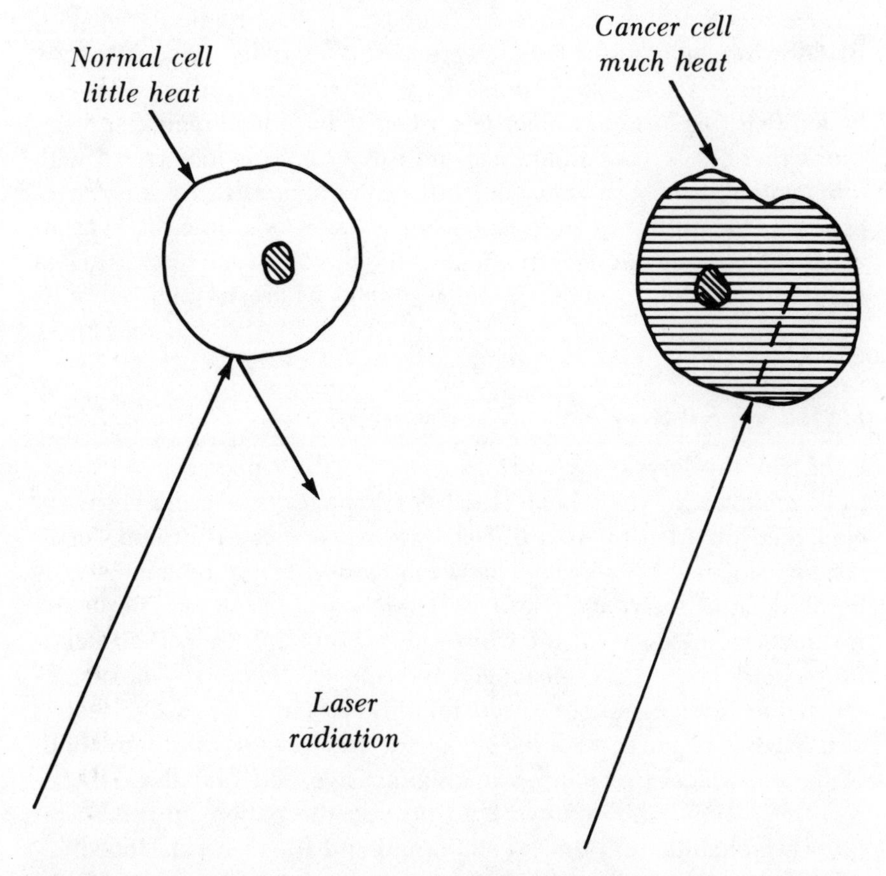

Fig. 3-5. Certain dyes cause cancer cells to absorb heat from a laser, and also light energy, triggering reactions that destroy the cancer cells but do not affect normal cells.

cause destruction of the malignant cells. And heating is not the only factor. When the dye breaks down because of the laser light, molecular fragments are produced, which may cause interference with the functioning of the cancer cells, preventing them from reproducing and even killing them by starvation. This method of cancer treatment is still largely in the developmental stage.

Even in cases deemed terminal, tumors have shrunk so that they can be removed by conventional surgery following laser radiation treatment. The ruby laser is often used with the dye, which is called HPD, or hematoporphyrin. The laser may be directed at the malignancy by means of an optical fiber, so that the radiation treatment can be performed without the need for surgery to gain access to the tumor.

Laser radiation may someday supersede other methods of malignant tumor treatment. Noncancerous tumors may also be destroyed in this way. Lung cancer has been treated at the Mayo Clinic by Drs. Denis Cortese and Robert Fantana in an outpatient capacity. Bladder cancer has also been treated by laser radiation at Mayo by Dr. Ralph Benson. The process takes about 45 minutes. It is believed that this form of cancer treatment may reduce recurrence because there is less "shedding" of the cancer cells as compared with other methods. There is no bleeding, general anesthesia is not necessary, and a catheter does not have to be used. Since this treatment can be done on an outpatient basis, the cost is significantly less than with conventional surgical operations, where hospitalization is necessary.

LARYNX SURGERY

Cancer of the larynx has, until recently, made it necessary to surgically remove the voice box. This has left patients with no voice except that provided by an artificial, external device. However, small lesions can now be removed by vaporization using a laser. This is being done at the Mayo Clinic by Dr. Nicolas Maragos. The larynx is undamaged except for the cancerous lesion, and the patient keeps his or her voice largely unchanged and intact. The carbon-dioxide laser is the most commonly used for this surgery. The CO_2 laser is also used to remove small lesions in the trachea, or windpipe. Both of these surgical procedures are noninvasive and bloodless. Treatment can usually be done without putting the patient in the hospital. The rehabilitation time is shortened and the cost is reduced.

The patient is first placed under general anesthesia. An endoscope is inserted, if necessary, into the airway. The laser light is beamed through the endoscope at the specific target area. Everyone in the operating area wears safety goggles because of the intensity of the laser light. The patient wears gauze pads over the eyes. While accidents are not likely, the intensity of the laser light is comparable to that of full sunshine at midday, and precautions must be taken.

LASER ACUPUNCTURE

In the Orient, needles inserted in various points on the skin have long been used for relieving pain and for producing certain physiological effects. In the West, this technique is not well understood and is not widely used. Experience has shown, however, that applying localized intense pressure at certain places on the body will cause very specific and predictable things to happen, often someplace not close to the pressure point.

Some examples of pressure points and the resulting physiological effects are; the center of the chest relieves pain of gastric origin, the backs of the shoulders relieves liver problems, the forehead relieves constipation.

Traditionally, acupuncture has been done using needles inserted into the pressure points. Perhaps you have seen a person sitting demurely with several needles sticking conspicuously from various unseemly places on the body, appearing mildly amused and looking around with an air of sophistication and glamour while vaguely resembling a porcupine. There must be something to the practice if a patient will put up with that. Acupuncture is not entirely without risk: the main precaution is that the needles must be sterile to prevent hepatitis or other infection.

A less painful and, in many cases, equally effective way to perform acupuncture is to use a small laser in place of a needle. This is done quite extensively in China, and is gaining popularity in Western Europe as well. Most of the laser acupuncture is done in the more well-equipped medical centers of large cities. Some practitioners think that lasers work more effectively than the conventional needles. The laser certainly presents less risk of infection. In Western Europe there are more than 2,000 laser acupuncture systems in current use.

An example of laser acupuncture is the application of laser light to pressure points located on the toes. This practice has been found to cause a fetus to rotate into the proper position prior to birth. This occurs in about four out of five cases in the last three months of a pregnancy. The success rate using lasers is higher than that obtained with conventional needles. Why this works at all is not well understood, but the fact remains that the phenomenon does occur and can be put to good use. It has been said that "one experimenter can keep a dozen theorists busy."

Among other conditions treated with laser acupuncture are bronchitis, asthma, and other chronic conditions involving pain or discomfort. In Europe, lasers are now used in about one out of three acupuncture treatments; the percentage is higher for children and nervous people. There is basically no sensation—not even warmth—from the low-powered lasers.

The medical establishment in the United States recognizes that this field is susceptible to exploitation by quacks, and remains cautiously skeptical, although there is some recognition of the practice and its use appears to be increasing.

CARDIOVASCULAR DISEASE

Until fairly recently, laser technology has not been used in the treatment of cardiovascular disease, mainly because of concern that healthy tissue might be damaged. The idea was clear enough: lasers might be employed to remove obstructions such as blood clots and arterial plaque. The main fear was accidental perforation of the blood vessel wall. In the past several years however, better understanding of potential problems, and how to best avoid them, has been gained.

A mass of arterial plaque is yellowish-white in color. Thus it was thought that a laser demonstrating maximum absorption by this color of material might be useful in boring a hole through an occlusion. Evidence indicated that using light of the correct wavelength would present little risk of damage to the arteries themselves, while allowing destruction of the plaque. The plaque material could also be stained, so that it would readily absorb laser light of a certain wavelength, to which the blood vessel wall would be relatively inert. Special fiber-bundle transmission devices were designed, along with metal-tipped fibers that result in heat rather than light emanating from the end of the fiber. The use of a metallic laser-heated probe is illustrated in FIG. 3-6.

Most arterial procedures using lasers have been done in the large vessels of the legs. Treatment of coronary-artery occlusion is still largely in the develomental stages. The coronary arteries are the ones that supply blood to the heart muscle. It is blockage of these arteries that leads to heart attacks. Various ideas have been put forth as to the best types of devices for use in the coronary arteries. A laser has even been considered that would allow the heart muscle to

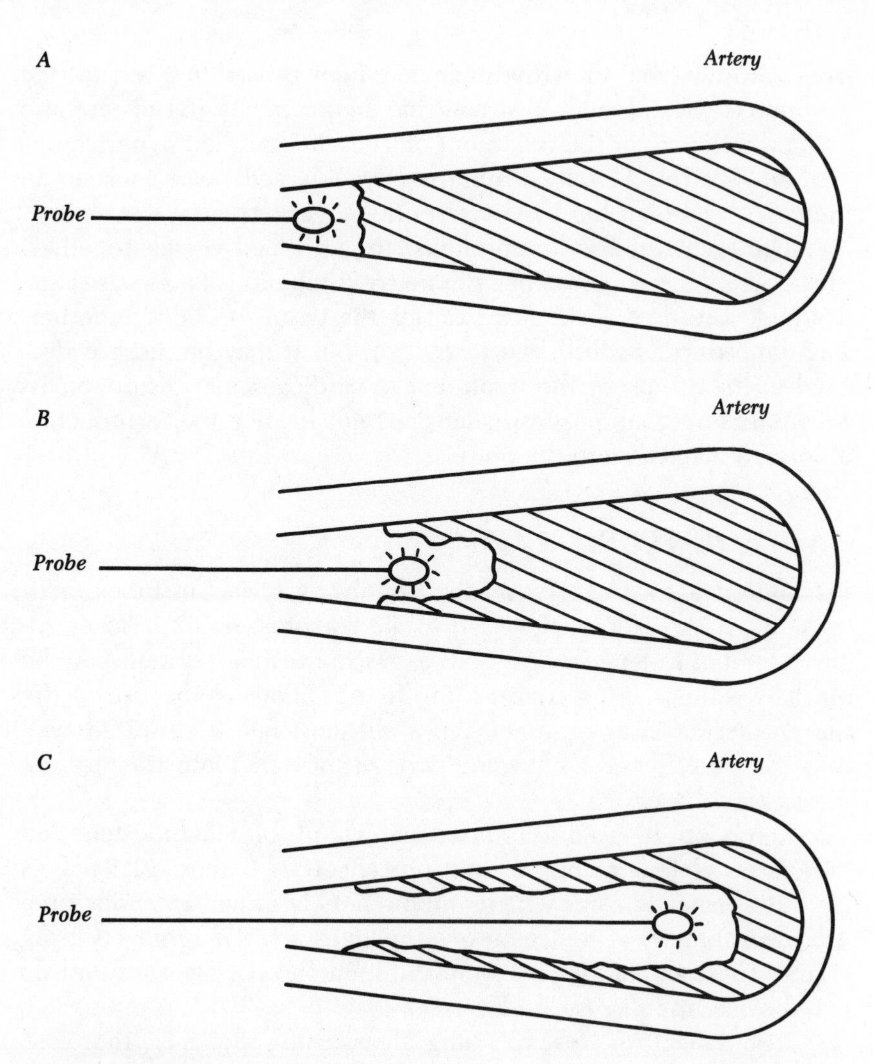

Fig. 3-6. A heat-tipped probe, heated by a laser, removes an arterial plaque. Progressive stages are shown at A, B, and C.

get its blood directly from the heart chamber. To accomplish this, it would be necessary to make small holes or "channels" in the heart itself, running from the outer surface directly into the ventricle. The problem in the past, when these channels have been made using conventional surgery, has been that they tend to close in time. However, when the holes are made using the laser, they seem to last longer.

The removal of plaque using a laser is known as direct laser revascularization. It is currently being tried on human patients.

The laser is being considered in further applications for the treatment of cardiovascular disease. One example is the treatment of irregular heartbeat, or arrhythmia. The laser is used to interrupt the conductive tissue bundle that controls the heart rate. Lasers are also being developed for the treatment of a condition called hypertrophic cardiomyopathy. This is an abnormally enlarged heart muscle. In the past, treatment has been conventional surgery.

The laser can also be employed to join blood vessels together. This operation, which has been done by stitching, is known as tissue welding. The heat of the laser causes the tissues to bond together. This procedure is still in the test stage, but it may be more widely used in the future for the treatment of cardiovascular disorders. Its advantages over sutures is that it does not involve the introduction of foreign material into the body.

GASTROENTEROLOGY

The digestive tract can be reached easily by means of tubes or fibers, and is therefore well suited to the use of lasers to perform minor operations. Some parts of the gastrointestinal system can be reached without using transmitting fibers, allowing the use of the carbon-dioxide laser or other types that are not transmitted well through these fibers. Only some parts of the small intestine are not accessible to lasers.

Lasers can be used to remove many kinds of benign lesions and tumors, as well as malignant and precancerous tissues. Flat polyps are often removed using a neodymium-YAG laser and an endoscope. Advanced tumors of the colon or esophagus are also removed using lasers. These operations are done to open the passageways and do not alleviate the disease itself.

It is difficult to treat bleeding ulcers using the laser, therefore the laser techniques is not often used in such cases. But a laser can be used if conventional surgery is considered too risky, or if no other method of treatment is deemed feasible.

Vascular lesions, similar to "port wine" skin discolorations, sometimes develop in the gastrointestinal tract. If they bleed, the laser is especially useful in treating them. Swollen veins in the esophagus, known as *varices*, can be made to shrink and coagulate by means of the laser. Varices develop in conjunction with cirrhosis of the liver, as a result of the body attempting to bypass an inefficient liver by building new blood vessels around it. Drugs or surgery are sometimes too risky for use in the case of bleeding varices.

New developments in gastrointestinal laser surgery include the use of sapphire contact tips. These tips may reduce the chance of tissue damage from the invading optical fibers. Experimentation has shown that a pulsed neodymium-YAG laser can be employed to break up gallstones. Efforts are currently underway to develop an endoscope that will safely deliver the laser light to the bile duct without affecting nearby tissues.

DERMATOLOGY AND PLASTIC SURGERY

Skin lesions were among the first medical conditions considered for treatment using the laser. This is not surprising since the skin is easily accessible and many kinds of lesions have in the past been literally "burned off" by direct application of heat, cold, or by cauterization.

The earliest application of the laser in dermatology was in the removal of so-called "port wine" stains, which get their name from their characteristic color. These stains occur in places where there are too many blood vessels in the skin. They are most often found on the head, neck, hands and arms, where they are often visible and can cause a person to be self-conscious. In some kinds of "port wine" stains, much or most of the discoloration can be removed by means of irradiation by a laser. The procedure does not require much time and is essentially painless. In many cases there is practically no visible stain or scarring after successful treatment. Other kinds of skin abnormalities that may be treated by means of the laser are hemangiomas other than the "port wine" type, teleangiectasias and "spider veins," which appear as weblike patterns. The reddening of the nose that sometimes follows plastic surgery may also be alleviated using the laser. Carbon-dioxide lasers have been employed to remove tatoos.

For vascular skin lesions, the argon laser, which is well absorbed by skin pigments and blood cells, has traditionally been used. The laser light is pulsed in bursts lasting about 0.3 milliseconds. The carbon-dioxide laser is also being used for destruction of vascular lesions. The laser beam is pulsed, rather than delivered continuously, to reduce the spread of heat from the beam through the skin, and to make scarring less likely.

The laser is used to treat various kinds of skin cancer, including melanoma in the early stages of malignancy. The melanoma, or mole-type skin cancer, is the most dangerous since it can spread to other parts of the body and may be fatal if not treated early enough. The laser method of treatment has not clearly proven itself better than the older means of skin cancer treatment, except in some forms of the disease. One idea being considered is the injection of special light-absorbing dyes into skin cancers, using certain antibodies that tend to be attracted to malignant cells. This would enhance the destruction of cancer cells and reduce the risk of damage to surrounding healthy cells.

NEUROSURGERY

The laser has been used in the treatment of some disorders of the nervous system. This was first done in the latter part of the 1970s and is now categorized in two ways—benign brain tumors and benign or malignant tumors in the spinal cord.

In the brain, near the base of the skull, tumors may develop that interfere with many brain functions. Surgical instruments are used to get to the tumor in the conventional way. Carbon-dioxide lasers may then be used to assist and to remove lesions without affecting surrounding tissue. This precision is especially important in brain surgery. Side effects of the surgery, which have in the past included various degradation in function of the central nervous system, are minimal. The recovery time is much shorter, which saves money and makes more hospital space available for other patients.

Lasers and traditional surgery are also being used together to treat and remove benign and malignant spinal tumors. As with brain surgery, the laser in spinal surgery has less effect on surrounding tissues and the rate of complications is less than with other surgical methods. For example, paralysis is less likely to result from laser operations on the spinal cord as compared to conventional surgery. While complete cures for malignant tumors are not much more likely, if at all, the trauma and recuperation times are greatly reduced when lasers are used.

OTOLARYNGOLOGY

The windpipe, lungs and larynx are readily accessible using endoscopes, and these parts of the body therefore lend themselves well to surgery using laser technology. The usual laser used in such treatments is the carbon-dioxide laser. The laser light is transmitted through a special rigid type of endoscope used in the head, neck and upper chest, since a more flexible endoscope will not readily transmit the light from this type of laser. The windpipe and larger bronchial tubes can be reached for surgical purposes using the more rigid devices.

Cancer of the vocal cords is fairly common and is often treated using lasers. Radiation therapy can sometimes be employed to shrivel up smaller lesions, but this requires exposure on alternate days for up to three months. Also, if the tumor reappears, the radiation treatment cannot be repeated. A recurrence in such a case may require conventional surgery to remove part or all of the voice box. However, a laser can remove small tumors completely, and if some of a lesion is missed on the first surgery, the procedure can be repeated as many times as necessary to destroy it all. The operation can be completed in about an hour and the stay in the hospital is rarely more than 24 hours. Physicians now believe that if the laser is universally employed for the removal of small tumors in the voice box, complete removal may be reduced by about 20 percent in the immediate future.

Basal-cell carcinomas of the ears have been treated by means of lasers. Even quite large tumors in the upper windpipe or esophagus have been removed using the laser. A complete cure is seldom obtained for especially large tumors, but the quality of the patient's life may be made better by making it easier to breathe or swallow.

A benign growth known as a *papilloma* is often treated by means of the laser. These tumors may grow in the nose, throat or other regions of the head and neck. These growths are caused by a virus and tend to recur often. It is thought that treatment via laser results in a lower reappearance rate than other, conventional methods.

Tonsillectomy is not performed today as often as it was a few years ago but removal of the tonsils is still sometimes deemed necessary. The laser can be used for this operation, and can also be used to treat chronic infections of the tonsils without actually removing them.

Ear infections can be treated using lasers. A tiny hole or cut is made in the eardrum, which allows the fluid from the middle ear to drain, relieving the pressure. In the past, this was done by manual means and often caused exquisite pain. The laser process is sterile, relatively painless and essentially bloodless, since the laser tends to coagulate the capillaries in the eardrum.

A wide variety of congential abnormalities of the head and neck region have been effectively treated using the laser. These include blockages such as "webs," abnormal buildups of blood vessels, and stenoses.

PULMONARY MEDICINE

The laser is rapidly gaining acceptance in the treatment of disorders of the windpipe and bronchial tubes, and it is being employed increasingly to remove malignant blockages of the trachea or bronchi.

The laser of choice in this application is the neodymium-YAG type since it is very effective for tumor removal, produces very little bleeding, and can be readily transmitted through a bronchoscope. One method of treatment uses both surgical removal and laser radiation therapy to remove a blocking tumor and then treat the remaining cancer to minimize new growth and prolong or eliminate the recurrence time. The absorbing dye, HPD, that collects in malignancies may be used to enchance laser radiation treatment by maximizing absorption of the laser light by cancer cells while not affecting the normal cells.

Benign tumors of the trachea and bronchi are sometimes treated with the carbon-dioxide laser if they are within range of the nonflexible endoscopes required for the transmission of this type of laser light. The endoscope must be inserted through the mouth and into the throat and windpipe from there, compared with the conventional bronchoscope which is usually brought in through the nose. These types of tumors can be treated if they are relatively small and have not affected the cartilage in the trachea. Tracheal blockages have also been treated by using the carbon-dioxide laser.

UROLOGY

Another field of medicine in which the laser is gaining wider application is in the treatment of conditions of the urinary system. Infectious warts, known as condylomas, occur on the surface tissues of the penis or vagina, and also at times on the insides of these organs. The recurrence rates are high for these types of lesions; however, the laser is now being used to treat the lesion areas following removal, making recurrence less likely. This process is called "flashing." The treatment destroys most or all of the latent viruses that may remain following removal of the warts themselves.

Urinary tract tumors have been treated using neodymium-YAG lasers. Bladder tumors called *transition-cell carcinomas* (FIG. 3-7) are removed using electrosurgery in the traditional method, but recurrence is frequent, even though the cancer does not usually spread rapidly. However, if an endoscope is used to direct the laser light at the tumor, it appears to reduce the rate of recurrence for this type of malignancy.

More advanced bladder tumors have been treated using the laser in conjunction with electrosurgery. The tumor itself, or most of it, is removed using electrosurgery, then the laser is directed at the remaining cancerous tissue to destroy whatever may be left. HPD

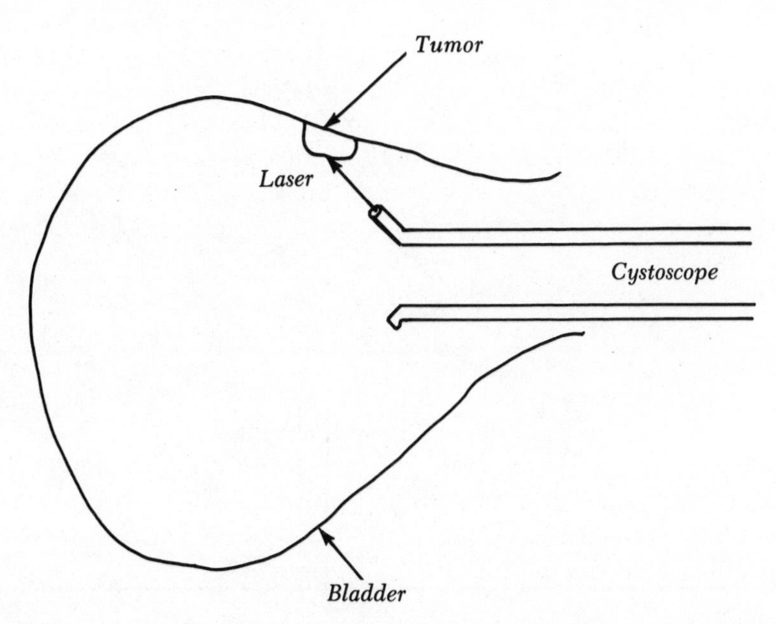

Fig. 3-7. A cystoscope, with a laser attachment, can be used to remove small cancerous bladder tumors.

dye may be employed to enhance the absorption of the neodymium-YAG laser light by the cancer cells while not affecting normal cells.

OTHER SURGICAL APPLICATIONS

Most operations of a general nature are still done in the conventional manner, using scalpels. The main reason for this is that laser consoles are so large it is difficult for them to be used in conjunction with a team of surgeons. Lung and breast cancer removal, for example, must usually be done in the older way because the laser console is simply too unwieldy.

The laser may someday be the tool of choice in most operations, however, and there are reasons besides simple extrapolation of technological advances to believe this. A good example is the general operation for cancer. Lasers provide a clear view, no contact between the instrument and the body, self-sterilization, less chance of complications, shorter healing time, and less bleeding. A scalpel causes bleeding from cut capillaries (FIG. 3-8A) which makes details hard to see. The laser cut is dry (FIG. 3-8B) and visibility is good even under low magnification.

This clear visibility, even using magnification of up to perhaps 40 times, allows greater precision using laser surgery than could be possible using conventional techniques. Tiny areas can be treated

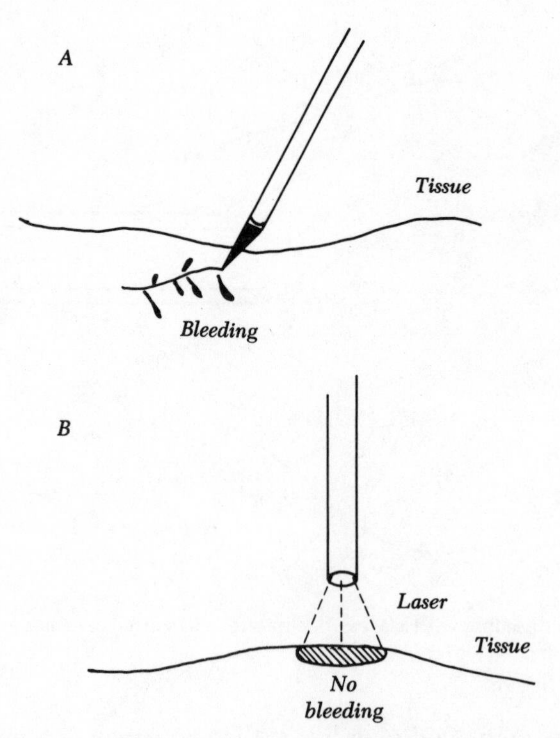

Fig. 3-8. (A) traditional scalpel causes bleeding. At B, laser, with no tissue contact, cauterizes its own cuts and there is usually no bleeding.

without affecting surrounding tissue. In the case of cancer, the laser may eliminate the need for cutting out large regions surrounding a tumor in order to be sure of getting all the cancer cells out.

The precision of the laser allows a surgeon to remove much smaller malignancies than is possible with conventional methods. Using no magnification, the surgeon can discern an area of about 100 million cells, or sometimes even a billion or 10 billion cells if the cells are especially small. Using magnification of 16 to 25 times, the number of cells in the view field is reduced to between 100,000 and as few as 100. With magnification of 400 times, as few as 10 cells may be isolated. The number of discernible cells is inversely proportional to the square of the magnification for a particular tissue sample.

The laser provides these advantages for general surgery: a dry field, complete removal, minimal trauma, minimal side effects, less post-operative swelling, less post-operative pain, sealing of blood and lymph vessels, ease of transmission using optical fibers if necessary, precision in removing malignant cells, repeatability, minimal mechanical effect in removing cancer cells, ease of sterilization, and a reduction in the need for transfusions. All of these advantages present a good case for any surgeon.

The laser has some promise in treating many conditions besides cancer. These may include dispersed infections. It will be necessary to design laser devices so that their use is not disruptive in the operating room. It will be necessary to be able to switch types of lasers quickly and easily.

LASERS AGAINST AIDS

One of the most promising developments for future use of the laser in medicine is in the battle against Acquired Immune Deficiency Syndrome (AIDS). Hardly a day passes in the life of the average citizen without at least one mention of this disease, already sounding such alarms in society. Certainly, unless a vaccine is soon found, this disease will become a modern-day plague as terrible as the plagues of ancient and medieval times. There is some hope developing, however.

Laser light can be used in conjunction with special dyes to eradicate cancer cells, and recently, this procedure has also been found effective in to killing a variety of viruses in the blood. The recent research has been done in Texas. Dr. James Matthews of Baylor University Medical Center has conducted laboratory experiments in this field and he reports that it appears blood for transfusions might be "cleansed" so it could be guaranteed free of AIDS and other deadly viruses. The flow rate would be approximately one pint of blood every 15 minutes or one liter every half hour. This would be a sufficient rate to make the method practical for routine use by blood banks.

Transfusions have been a cause of AIDS, and the recent testing, while eliminating much of the contaminated blood, is not entirely foolproof. The laser "cleansing" process would provide additional safeguarding of the blood supply.

The work by Matthews and his colleagues was partly financed by the Pentagon in conjunction with the Strategic Defense Initiative (SDI, or "Star Wars") anti-missile program. The first paper concerning the laser technique for blood cleansing was published in *Transfusion*, a journal of the American Association of Blood Banks. The laser technique appears not to harm the blood, leaving healthy cells alone, but has devastating effects on pathologic viruses, of which AIDS is only one.

The herpes simplex virus has been destroyed in blood using the laser and dye techniques. Flowing culture medium tests, but not blood tests, have been done with the AIDS virus, and this virus, too, was killed without harm to healthy tissues or cells. It will probably take three to five years from the time of this writing to perfect the technique before it is in routine use by blood banks. The procedure has been under development since 1984 and may be in widespread use by 1991 or 1992.

The AIDS virus is, of course, transmitted by sexual contact and by infected, shared needles as used by those who "mainline" street drugs. The laser technique is not a cure for the disease, and its control is largely a matter of personal conduct. It may be possible that the laser method of blood cleansing now under development will someday provide some insight as to how the virus may be eradicated from an entire human organism, but that is at least several years away.

The laser treatment of blood is especially significant since tainted blood may not test positive for several years after initial exposure.

Ironically, Star Wars funding will be crucial in the effort to perfect this new blood-cleansing system, since it will allow medical researchers to work with free-electron lasers, which are being employed in SDI.

THE FUTURE USE OF LASERS IN MEDICINE

Lasers will almost certainly gain acceptance in various medical applications as technology improves and laser consoles are made easier to use in operating rooms. In general surgery, the applications are almost unlimited, and the laser may even make it possible to perform new kinds of operations never attempted by conventional means.

Lasers are used in instruments such as cell sorters, flow meters, spectrometry and other devices, and it can be expected that their applications in these fields will grow. An interesting possibility is a microscopic hologram that may be made using lasers.

Extremely small-diameter lasers would be capable of modifying cells individually. It might even be possible to change the chromosome configuration or DNA in a cell, perhaps affecting those cells that reproduce from it. So *genetic engineering* might make use of lasers. This field is a controversial one, since there are those who fear the use of genetic engineering to attain political ends. We can be assured that it will be a long time before scientists agree to make mass sperm for modified eggs that will produce good soldiers—but then, just a hundred years ago, the idea that man would set foot on the moon was confined to imaginations such as Jules Verne's. Anything can be tried if it is theoretically possible.

Lasers are commonly used for the alignment of various instruments, and for proper alignment of the body when using such devices as CAT scanners or NMRI machines. I recall two occasions on which I was put through a CAT scanner—a device for taking multiple cross-sectional X-ray photographs to look for such things as blood clots in an arm or leg. There were several small red laser apertures with warning signs that admonished me not to "stare into the beam." I wondered who would stare into a painfully bright light, but the warnings were sufficient to keep me from looking to find out. The attendant said that the lasers were simply to let the technician check to see that my body was positioned correctly when passing through the machine. It had a futuristic appearance giving the impression that on the other end, I might come out in a Roman chariot race or on board an intergalactic cruiser. Then the procedure was over—no bizarre effects had taken place on account of the lasers. What a letdown.

Lasers in medicine are as common as any other instruments, and may already be the most widely used of any single tool in this field.

CAREERS IN LASER MEDICINE

Laser medicine is a fairly new field and has many different possibilities, so careers are not yet very well-defined. Training as a medical technician or nurse is probably the best route for someone interested in working with lasers in the medical field. But the more obvious goal would be to become a surgeon specializing in areas where laser use is widespread now, and is likely to remain so in the future. Training on the job has been the accepted way in which laser specialists have been developed to date. Current careers in laser medicine are the Laser Nurse Specialist and the Biolaser Technician.

The Laser Nurse Specialist is usually a graduate nurse. Their responsibilities include telling the patient about the kind of treatment that is planned, and monitoring the operation of the laser apparatus. A knowledge of laser physics is needed, along with a working knowledge of the various types of lasers and their characteristics and medical applications. After the laser procedure, the Laser Nurse Specialist evaluates the patient for discharge from the hospital.

Numerous hospitals offer short courses for nurses in the field of laser medicine. Nurses may also take some courses for physicians in the field.

The Biolaser Technician assists the doctor in operating the laser during a medical procedure, much as an X-ray technician operates the X-ray machine. The Biolaser Technician may also perform routine maintenance on the laser equipment. Some people train as general laser technicians and later specialize. Some get their medical training on the job. Much of the Biolaser Technician's knowledge is gained by experience.

Many clinics and hospitals employ people to make sure that laser safety standards are followed. This includes the wearing of special goggles by all personnel using the laser.

It is expected that opportunities for employment in the laser medical field will continue to increase as the use of these devices in medicine becomes more and more widespread.

4

Lasers in Communications

Lasers are well suited for propagation over long distances because their beams do not spread nearly as much as beams of light from conventional sources. Generally, light from a point source, or from any relatively distant source, becomes less intense according to the square of the distance. That is, if d_1 and d_2 are distances from a source and L_1 and L_2 are the intensities as observed from these distances, respectively, and all measurements are in the same units, then

$$\frac{L_2}{L_1} = \frac{d_1^2}{d_2^2}$$

This "inverse-square law" is illustrated in FIG. 4-1.

The inverse-square law for the intensity of light applies to light that is collimated, or beamed, just as it does to a light source that is a tiny point. At a certain distance from a beamed light source, the beam can no longer be considered as consisting of parallel rays of light, but is made up instead of diverging rays. This spreading would not exist in theory if the light source were a true geometric point, with no dimensional measurements, as shown in the reflector example of FIG. 4-2A. The rays of light would remain parallel forever as they traveled away from the source. But in the real case (FIG. 4-2B), the source is not a perfect point and it casts a magnified image at great distances. This image gets larger as the distance is increased, doubling in height and width as the distance is doubled. Therefore,

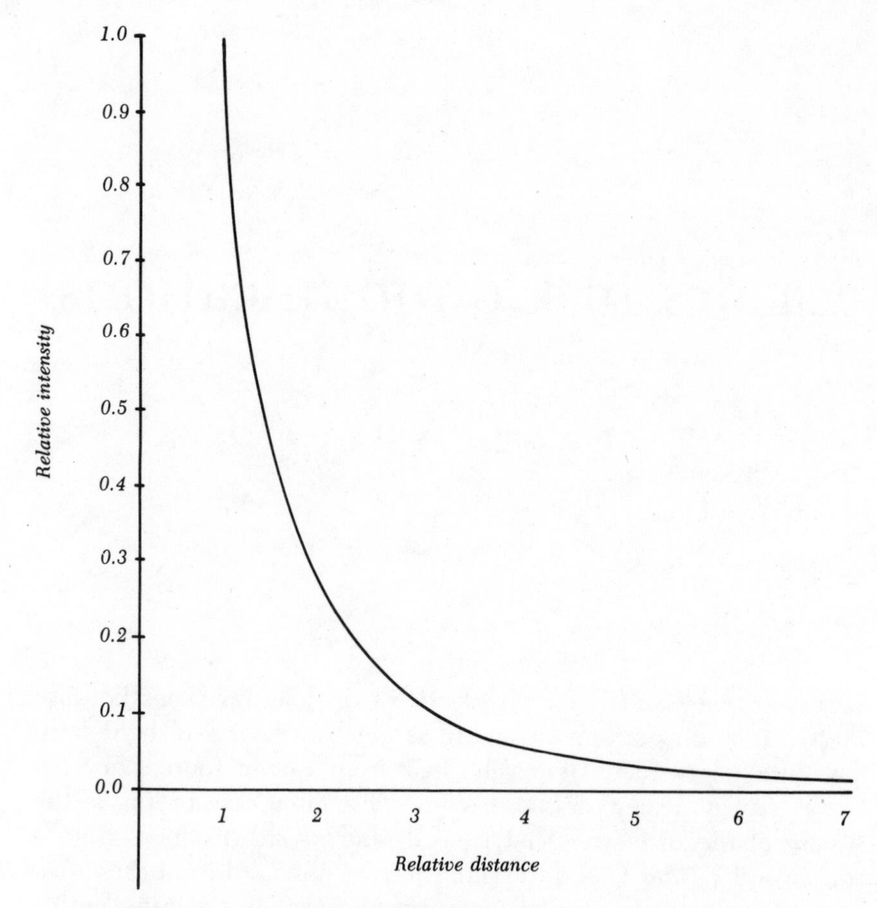

Fig. 4-1. Inverse-square law for light intensity versus distance.

if the distance is multiplied by n, the area of the image grows by n^2 while the amount of energy in the image remains the same. If a receptor is small compared to the image, increasing the distance to the source causes the intensity to diminish according to the same inverse-square relation as would be the case if the source were a point without a reflector.

In the atmosphere, some light is absorbed by the air molecules, especially at certain wavelengths. This does not have much effect on white light with constituent wavelengths ranging from over 7500 angstroms to less than 3900 angstroms, although most of this absorption is at the shorter (blue and violet) wavelengths. Dust and water vapor, as well as various pollutants such as particulate matter, ozone and carbon monoxide, all increase the absorption of light by the atmosphere. Water droplets and particulate matter cause some scattering of the light as well. Absorption and scattering reduce attenua-

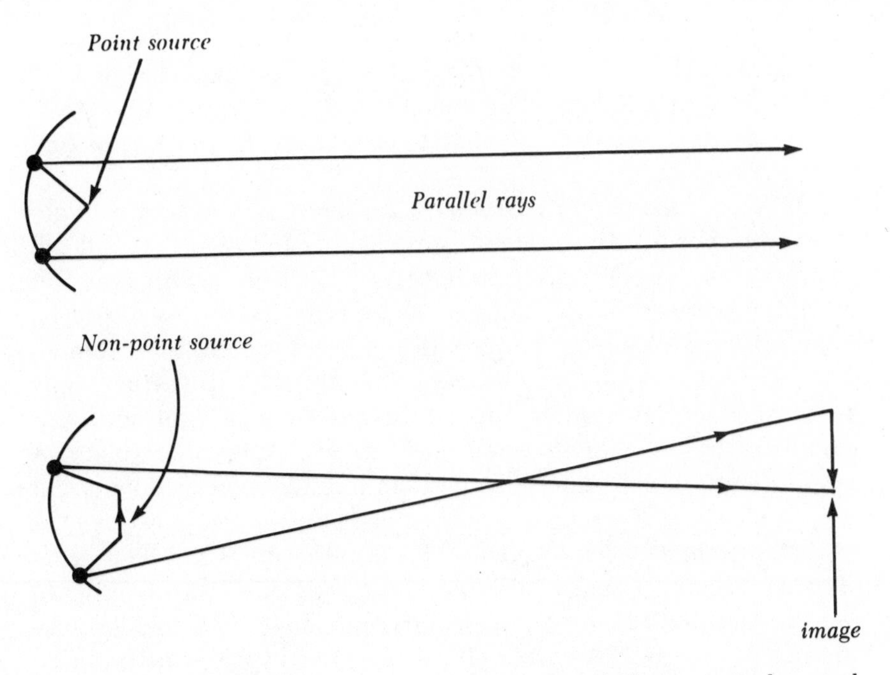

Fig. 4-2. At A, idealized point-source collimation of a light beam. At B, the actual case with a non-point source.

tion at a more rapid rate than would be the case with the inverse-square law in a perfect vacuum.

An ideal laser creates a beam of light with rays that are perfectly parallel. Therefore the intensity does not change, in theory, with increasing distance from the source. In practice, this ideal can only be approached, but cannot truly be reached, as is the case with the point-source collimator (FIG. 4-2A). However, in practice the laser comes much closer than any other type of light source to this ideal. A sophisticated laser concentrates its energy into a beam that remains narrow for many miles. This is why lasers may be used for destructive purposes—the "ray guns" of space warfare. It is also the reason lasers lend themselves well to line-of-sight, long-distance communications through space or the atmosphere. The inverse-square loss is defeated over large distances.

MODULATED LIGHT

In order to convey information, some characteristic of a light beam must vary. The most common way to impress information onto a beam of light is to vary its intensity. This is called *amplitude modulation* (AM) and resembles AM radio. We may also vary the wavelength of the light beam; this may be called *wavelength modulation* or *frequency modulation* or perhaps *color modulation*. The phase of the light beam, if it consists of a single wavelength, may be varied. The polarization, too, can be made to change for the purpose of conveying information.

Of these modulation techniques, AM is by far the easiest to obtain in practice. It can be done by switching the beam on and off, perhaps using fast-acting shutters (FIG. 4-3). This is still done for signalling between ships. Sunlight can be reflected from a mirror on a clear day; powerful arc lamps with reflectors can easily be aimed, seen, and read visually for distances limited, in good weather, only by the horizon. In a more sophisticated system, a light source is modulated by the audio-frequency energy from a voice amplifier. A simple technique for doing this with an incandescent or fluorescent lamp is shown at the schematic diagram of FIG. 4-4.

The modulated-light signal varies in intensity as if it were a radio wave of exceptionally high frequency and short wavelength. In fact the range of the visual wavelength from about 7500 to 3900 angstroms (7.5×10^{-7} to 3.9×10^{-7} mm) corresponds to frequencies of 4.0×10^{14} to 7.7×10^{14} hertz or 400 million to 770 million megahertz. These frequencies are so high that modulation can be had using modulating signals into the ultra-high-frequency range, and having very broad bandwidths. This is why it is possible to impress hundreds of thousands of different signals onto the same beam of light. The modulating signals may even be modulated themselves; for example, we may put television VHF Channel 5 onto a laser beam at 8400 angstroms. This would be an infrared gallium arsenide laser.

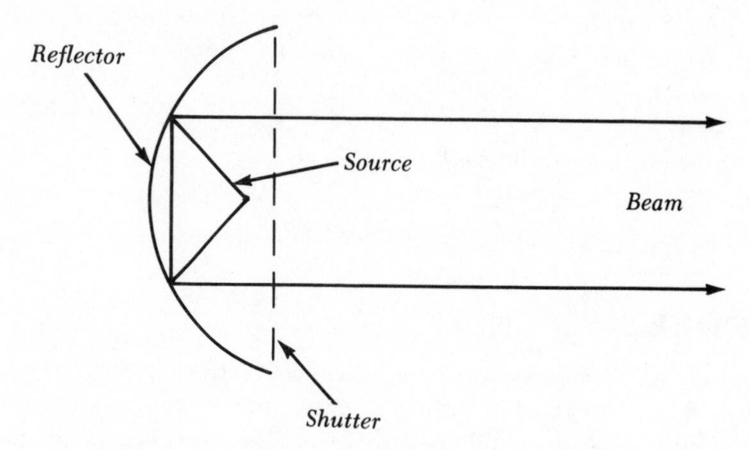

Fig. 4-3. A shutter can be used to modulate a light beam digitally at a slow rate.

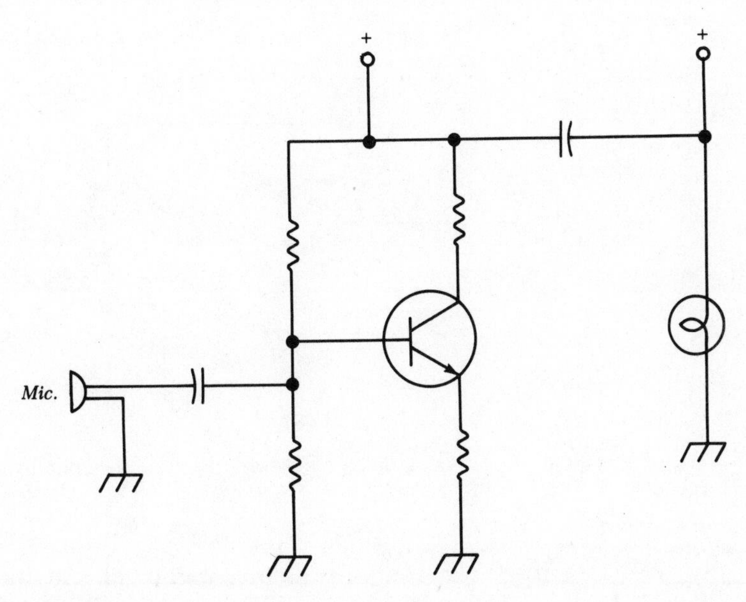

Fig. 4-4. Simple circuit for producing modulated light with an incandescent or fluorescent bulb.

Laser communication is limited to the line of sight, but this seemingly severe restriction can easily be overcome using well-placed reflecting devices or optical fibers. The latter medium is now beginning to replace wire systems in telephone networks.

Reception of a modulated-light signal involves the use of a photodiode, phototransistor or other light-sensitive device that is fast-acting enough to follow the modulation waveform of the signal sent. A simple circuit for detecting modulated light is illustrated in the schematic diagram of FIG. 4-5. The solar cell provides fluctuating direct current which is amplified by the transistor circuit. The audio variations are reproduced at the output. Further gain may be obtained by using a low-noise field-effect transistor stage, followed by several stages using bipolar transistors or an operational amplifier.

If the signal modulating the light beam is itself a modulated radio-frequency signal, then two stages of detection are needed; one to extract the radio signal from the beam of light, and another to extract the audio or other signal from the radio-frequency signal. In such a case there are two modulators at the transmitting station, as well. FIGURE 4-6 shows a block diagram of a complete one-way transmission system for sending Channel 5 (VHF) television signals via modulated light.

A modulated-light source suitable for use in the scheme of Fig. 4-6 would have to be capable of following changes in light-beam

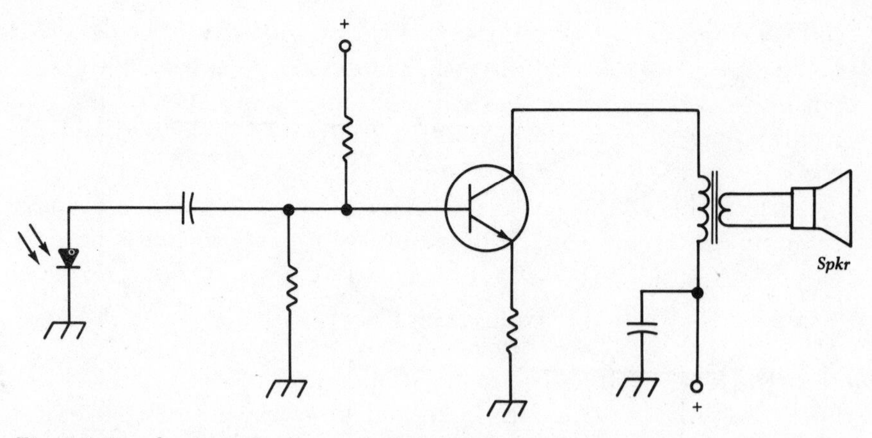

Fig. 4-5. Simple circuit for detecting modulated light. In this example a photovoltaic cell is used.

intensity well into the VHF radio-frequency spectrum. An incandescent light bulb is obviously not suitable in this application. Its filament can scarcely keep up with variations in audio frequency enough to convey any information at all. The laser is the ideal choice for this type of communications service. The power output (intensity) of the laser depends on the kind of communications desired. Long-range space communications would require a more powerful device, of course, than would a short-range circuit utilizing fiber optics.

THE FIRST PHOTOPHONE

The idea of modulated light is not new. In fact, the first communications system using this idea was designed in the late eighteenth century by the Frenchman Claude Chappe. He used towers atop hills,

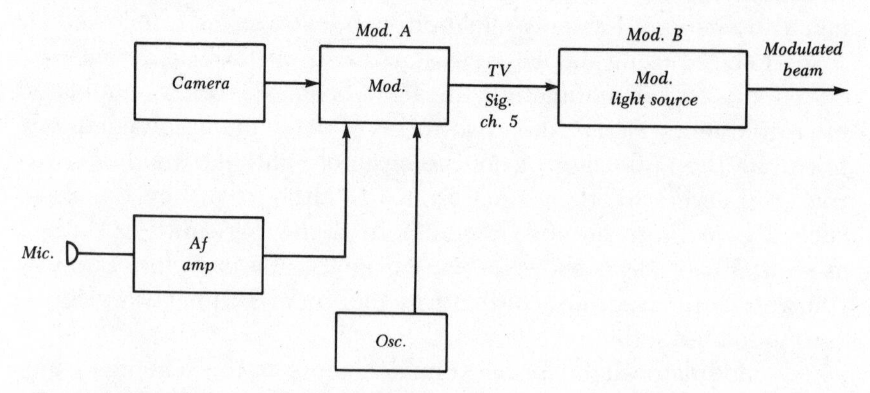

Fig. 4-6. Complete system for transmitting a television signal over a light beam.

and semaphores to indicate the various letters of the alphabet. This was about as crude a modulated-light system as can be imagined, but by relaying signals it was possible to send messages rapidly and accurately over ranges of thousands of miles when the weather was good.

Alexander Graham Bell was the next to use modulated light for communications. In 1880 he demonstrated the photophone. This device transmitted voice information over a beam of light reflected from a vibrating mirror. The mirror was used to reflect sunlight, and was made to vibrate with voice information. The signal was received by photosensitive devices similar to modern photocells. It worked, but its usefulness was severely limited. It would not function in fog or in cloudy weather, it would not work at night, and the range was limited, too, by the primitive hardware. The original photophone was put in the Smithsonian and virtually forgotten for many years.

The idea of the photophone was brought back when it was clear that an optical maser, or laser, could be developed. When, in 1960, the laser was invented and demonstrated, one of the first suggested applications was communications. Although communication by laser would be limited to the line of sight, the light beams would be extremely hard to "jam" because of their narrowness, and it was known even then that light beams would follow fibers of glass by being trapped within by total internal reflection.

DESIGN OF THE OPTICAL FIBER

Very thin fibers of glass demonstrate elasticity: They can be bent to some extent without breaking. Perhaps you have noticed the elastic properties of glass when a friend leaned against a large window pane and you saw the distortion of the reflected images. (Don't try this indiscriminately!) When an optical fiber is bent, the light within follows, since reflection occurs from the inside surface.

A modern optical fiber consists of a core with a high index of refraction, a surrounding material called "cladding" with a lower index of refraction, and an opaque jacket (FIG. 4-7). The beam of light is reflected internally at the boundary plane between the inner core and "cladding," provided that the angle of incidence between the boundary and the light beam is sufficiently small. If the angle of incidence is too large, the ray of light will pass through the boundary plane and is lost. However relatively little energy is lost in this way.

Using clear core glass and a light beam having a wavelength which glass transmits very well, propagation will occur for very long distances even with light beams of low intensity. Light can be

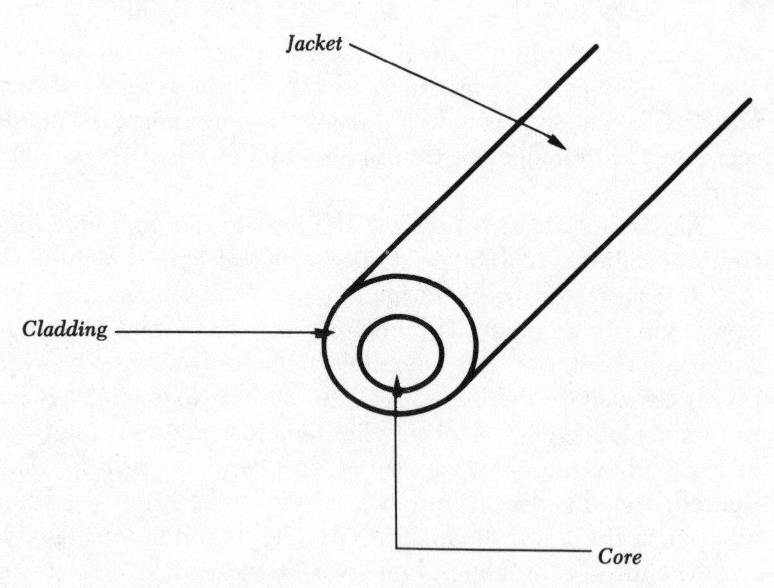

Fig. 4-7. Cross section of simple optical fiber cable.

passed around bends, very much like electromagnetic fields are propagated through a waveguide.

The first optical fibers were of low quality and the light was not generated by a laser. The poor transmittivity of the glass, and the incoherent and broadbanded nature of the light limited the range to a few meters. In 1970, the first long-distance optical fiber communication was done by Robert Maurer of Corning Glass Works. An especially pure grade of glass then became available from Standard Telecommunication Laboratories.

The transmittivity of an optical fiber can be expressed as the length of fiber in which the energy loss is 10 dB. This means that only 10 percent of the input energy reaches the output end. Around 1965, this distance was on the order of 10 meters, or about 30 feet. Maurer's low-loss fibers increased this distance to about 500 meters —more than a quarter of a mile. The best modern fibers have a 10-dB attenuation distance of about 20-25 kilometers, or 13-16 miles. These figures are for wavelengths where light is transmitted best. Modern optical fibers have the least attenuation in the infrared part of the spectrum at wavelengths of about 1.4 micrometers (1400 nanometers or 14,000 angstroms—about twice the wavelength for bright red visible light).

FIBER-OPTICS PRINCIPLES

The transmission of information using fiber optics is quite similar to radio-frequency energy sent via cable. A modulated light source feeds its beam into one end of the optical fiber or fiber bundle. The fiber or bundle is encased in flexible, opaque material and the whole system resembles an electric cable in appearance. The receiver at the opposite end senses and demodulates the light, and amplifies the component signals so they can drive a loudspeaker, picture tube, fascimile machine, telephone, or other device for retreiving the information.

A long-distance fiber-optics system must have *repeaters* at intervals along the length of the cable, just as radio-frequency cable systems. The repeaters consist of receivers that amplify the signals and retransmit them, usually, but not necessarily, using the same kind of light source as the transmitter. FIGURE 4-8 is a block diagram of a simple repeater, consisting of the photoelectric or photovoltaic cell, amplifier stages, power source, modulator, and laser.

Two types of light source are customarily used with a fiber-optics communications system: the laser semiconductor and the light-emitting diode. The laser semiconductor may be a diode or transistor, and the light-emitting diode (LED) may or may not have laser properties. The laser semiconductor produces a laser beam that spreads out much more rapidly, over a distance, than larger types of lasers. However, this is not a serious problem in fiber optics since the fiber keeps the beam confined. A collimating lens may be used at the input end of the fiber, keeping most or all of the light rays nearly parallel so that none of them escape the fiber to be absorbed by the jacket. The laser semiconductor or LED emits light when a sufficient current passes through it. The emitted light is proportional to this current and the rate of response is rapid, so that the light can readily be modulated by varying the current. It is not difficult to obtain 100 percent amplitude modulation in this way, so that the light available is used to maximum efficiency. This modulated beam passes along the optical fiber, so that the output intensity varies in the same proportion all along the fiber, even though the overall intensity decreases because of loss. The signal at the receiving end of the fiber is therefore identical in modulation waveshape to that at the transmitting end.

From the standpoint of receiving, it makes no difference whether the light source at the transmitter is a laser or not. The laser produces

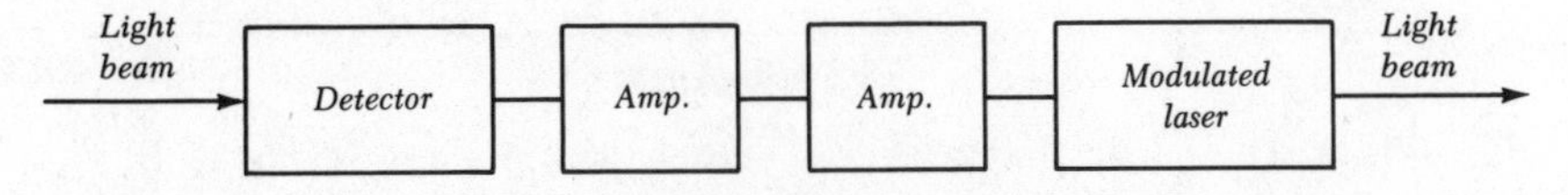

Fig. 4-8. Modulated-light repeater.

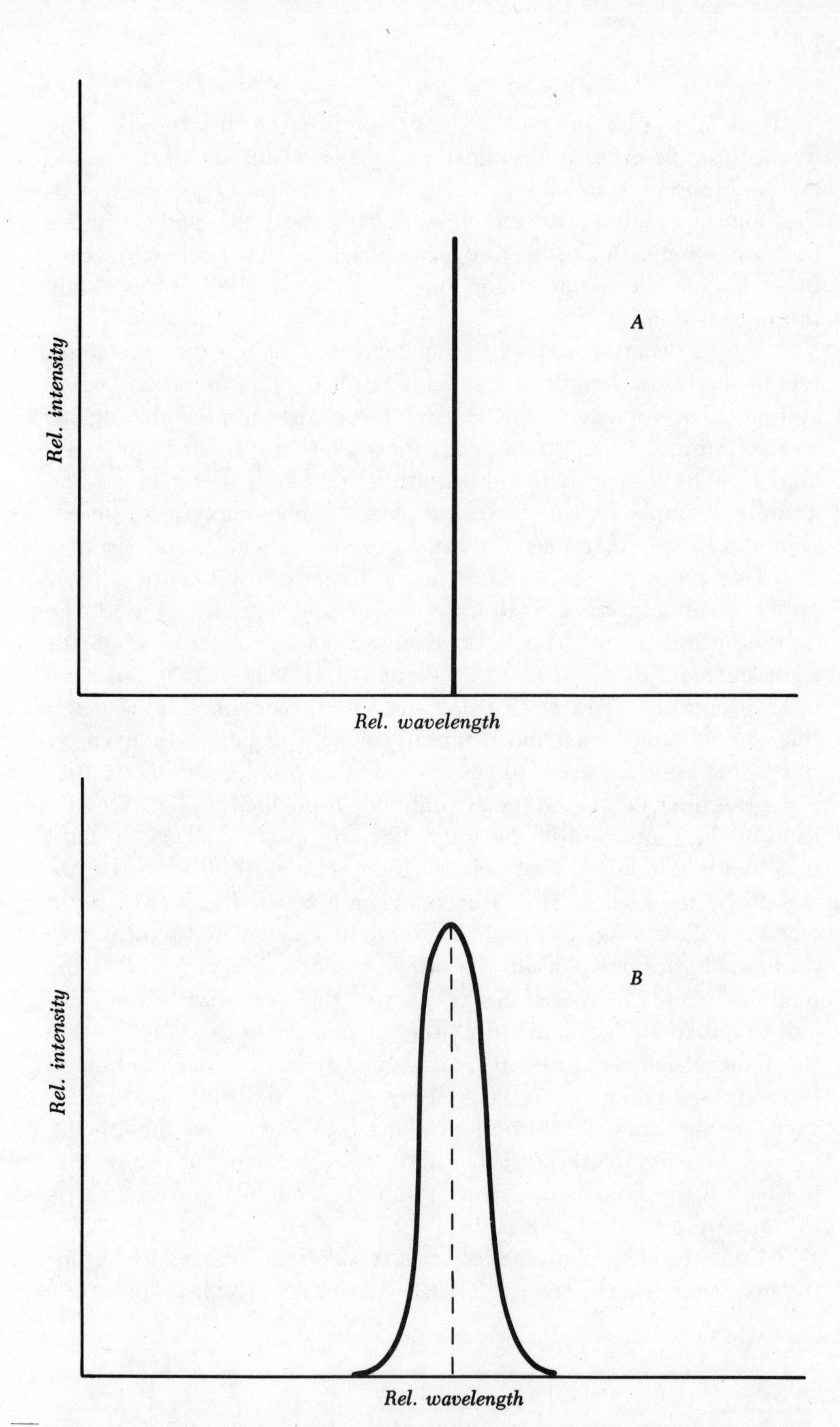

Fig. 4-9. (A) The output from a laser is entirely at a single wavelength. (B) The output from a light-emitting diode appears monochromatic but occupies some bandwidth.

a light beam having one discrete wavelength, while the LED produces light over a narrow but definable band of wavelengths (FIG. 4-9 A and B). As long as all the light is attenuated to the same extent as it travels along the optical fiber, it makes no difference if the wavelength distribution is imperfect. Also, the polarization is not of any importance and it isn't necessary that all the light waves be polarized in the same direction. In fact LEDs cost less and last longer in fiber-optics applications compared with lasers. Therefore the laser may be seen less and less in future fiber-optics developments. However, for space communications over vast distances in a vacuum, lasers will see an increasing role as space travel becomes more commonplace—and we as scientists surely look forward to that.

GETTING THE MOST INFORMATION ACROSS

A fiber-optics system can transmit much more information than a single cable carrying a radio-frequency signal. There are several reasons for this.

The first reason is that the frequency of light is much higher than the frequency of any radio signal. The maximum modulating frequency that a carrier signal can effectively convey is limited to about 10 percent of the carrier frequency. For a 10-MHz radio signal this limiting maximum is then 1 MHz. But for a carrier frequency of, say, 200 THz (1 THz = 1 terahertz = 1,000,000 MHz), the entire radio spectrum is easily within reach for modulation purposes. Actually, a carrier wave is rarely modulated at more than about 1 percent of its actual frequency, but the point is clear. Quite a few 3-kHz-wide single sideband channels can be impressed on an infrared beam having a frequency of 200 THz!

Another reason that modulated light is more effective than carrier transmission is that fewer repeaters are needed along a given circuit in an optical-fiber communications system, as compared with a wire system. The attenuation is less in a good, low-loss fiber system, and there is essentially no interference from outide sources.

Still another reason why modulated light is superior is that there is less trouble with *intermodulation distortion* (IMD) as compared with a carrier system. The IMD results from mixing of the modulating signals. For example, two modulating signals of 140 kHz and 180 kHz will mix to produce sum and difference frequencies of 320 kHz and 40 kHz. These signals are not intended to be on the carrier at all. They are "false" signals and they may be horribly distorted if both the original signals are modulated, say, by voice information. Nonetheless they will be as "real" to the receiver as if they were deliberately transmitted, although they will not be as strong as the original signals. These IMD products will cause interference with any actual signals we may want to convey at 40 kHz or 320 kHz. Furthermore, these IMD products may themselves mix with each other and with the original signals, producing still more interfering signals. If there are several hundred original signals at the transmitter, and there is much IMD, the result at the receiver will be a poor-grade quality of what might be called "goulash." It is quite difficult to prevent IMD. This is especially true in a radio-frequency system. There is less likely to be IMD in a properly designed optical communications system. Moreover, there is far more spectrum space available, so that modulating signals can be assigned frequencies that do not fall on, or produce, IMD with respect to other signals, while still allowing many hundreds of individual signals to be transmitted.

Digital signal modulation is easily accomplished using optical networks. Light-emitting diodes do not follow analog signals very well at high modulating frequencies. But digital modulation can be easily followed into the radio-frequency range by LEDs. Digital modulation has several advantages over analog, or continuous, modulation. The signal-to-noise ratio is superior, digital modulation requires less bandwidth, it is more easily handled by LEDs, and it is well suited for switching via computers. The digital signal is either all the way on, or all the way off. Thus it represents the most complete possible modulation of a carrier signal. Voices, music, and even video signals can be digitized. This allows far more information to be sent, using less power and with less interference, than would be possible using analog forms of modulation.

A digital voice modulated-light system is diagrammed in FIG. 4-10. The voice is first amplified and then changed by the analog-to-digital (A/D) converter into a series of 0s and 1s. The 0 state represents the complete absence of the carrier and the 1 state represents full carrier. This signal modulates the light beam so that the LED is

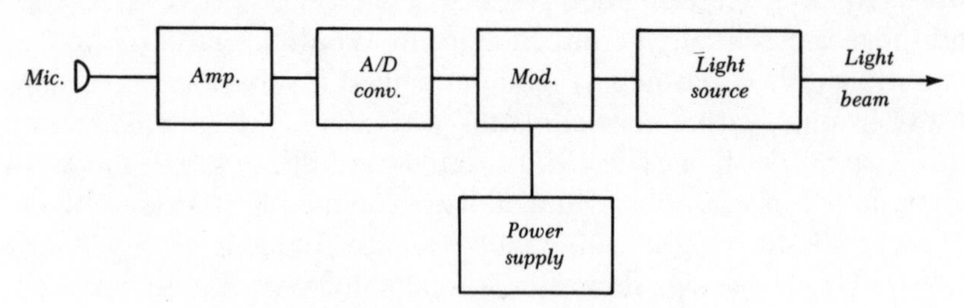

Fig. 4-10. Digital modulated-light transmitter for voice audio.

either full-on or full-off during any given instant of time. At the receiver, the detector is actually a digital-to-analog converter, or D/A converter. This retrieves the voice information. The receiving amplification may be done either prior to D/A conversion, or after the conversion. Amplification may also be acheived both before and after the conversion.

OPTICAL TELEPHONE TRUNK LINES

The optical-fiber method of signal transmission is ideally suited for use in telephone trunk lines. The trunk line is a main carrying line that handles many signals simultaneously. Traditionally, a trunk line has been an electrical cable with many separate conductors. However, all the signals that must be accommodated by the trunk line can be impressed on a single beam of light in most cases, and therefore a single glass fiber can take the place of a far bulkier and more expensive cable. The fiber is easier to maintain than the wire cable, it lasts longer, and it needs fewer repeaters. Also, as one company has put it, there isn't likely to be a shortage of the material—sand—needed for the manufacture of optical fibers.

Optical fibers can be used under the sea as well as on land. The glass is not subject to *intrusion* or corrosion as are wire cables. Undersea cables may take the place of microwave satellite telephone links someday. The optical system would cost less per hookup and the lag time would be practically eliminated. For a satellite in geostationary orbit, the radio waves must travel a round-trip distance of about 46,000 miles. This requires 0.25 seconds for completion of the circuit from transmitter to receiver and another 0.25 seconds for the reply. With an undersea cable the delay would be on the order of 0.07 seconds each way.

A metallic undersea cable requires one repeater every 9 kilometers or every 6 miles. With modern optical fibers, each carrying about 4,000 telephone conversations each, repeaters are needed only about half as frequently along the cable, and this reduces the chance of having to haul up several miles of cable to repair a malfunctioning repeater. Eventually the repeaters will be spaced at much larger intervals as fiber technology advances. Perhaps someday it will be necessary to have only one repeater for every 100 miles (160 kilometers).

In the event of a repeater malfunction, hauling up a fiber-bundle cable will be far easier than retreiving a metal cable, simply because the fiber cable is much lighter.

GENERAL APPLICATIONS FOR OPTICAL FIBERS

Optical fibers are also superior to wire cables in other uses besides telecommunications. Any kind of signal, such as a computer command, can be carried by means of fiber optics. This method of carrying signals is not subject to electromagnetic noise. Furthermore, it would be immune to damage caused by the electromagnetic pulse set up by a gigantic nuclear detonation at high altitude. This is an important defense consideration, since a few such explosions at strategic locations could cripple the whole country's communication system prior to an all-out nuclear war. While we may not like thinking about such an eventuality, and some might argue that being well prepared is just one step short of actually planning nuclear warfare, it would be foolish to insist on yesterday's technology solely because being vulnerable to annihilation might make us think twice before becoming involved in potential nuclear conflict.

Fiber-optics networks generate no sparks; hence there is no risk of explosion in volatile environments. The shock hazard, too, is eliminated. Damage from nearby lightning strikes, similar to that caused by a nuclear electromagnetic pulse but on a localized scale, would be greatly reduced or eliminated.

The fact that fiber-optics systems are immune to electrical noise interests automobile manufacturers. The microprocessors in modern automobile systems can be adversely affected by ignition noise and electromagnetic interference from mobile telephones and radio sets. Using fiber optics for transferring microcomputer commands would greatly reduce the risk of engine malfunction or the false indication that something is wrong. Moreover, there would be essentially no interference from the microcomputer that would degrade the performance of a mobile telephone or radio. A fiber-optics network would save weight and would not corrode in adverse weather. Far fewer cables would have to pass through the firewall to the dashboard, which would make servicing easier and would also increase reliability while minimizing fuel-consuming weight.

When a new technology is tested and developed in a device such as an automobile, where malfunction may carry the risk of injury or death, the nonessential systems are developed first, and the essential systems are developed only after the new technology has proven reliable. Fiber-optics systems will therefore be seen first in such applications as automatic windows and door locks, and later, when the "bugs" are out, in such things as electronic steering and brake systems. A fully fiber-optic controlled car may be two decades away as of the time this is being written. Other inventions may intervene in the meantime and we might never see a fully fiber-optic controlled automobile at all.

Fiber optics, by providing broad-channel communications, may be instrumental in the development of the video telephone. This long-dreamed-of (and nightmare-producing?) telephone is now well within the capability of technology. It is primarily a matter of installing fiber optics in the subscriber lines. Trunk lines for some telephone companies are already largely optical fibers. A video telephone could enhance computer interfacing. Home educational systems with complete two-way accessibility might, for example, bring school to every household. Shopping could be done entirely by video telephone. We might never have to leave the house again!

Such technology might be better suited to colonies on the Moon or Mars than to civilization on Earth, although there would in reality be many advantages to bringing jobs, school and services to the people rather than vice-versa. Transportation expenses would be reduced, pollution would decrease, and leaving the "cocoon" would be mainly a matter of recreational interest. Exercise and socializing would top the list of outside activities. Perhaps even the traditional city, as we know it, would become almost obsolete for a large part of the population.

THE FUTURE OF FIBER OPTICS

Eventually all the countries of the world may be linked by fiber-optics communications. Even now, it is within the scope of technology to call a friend living in India and converse at length by means of a videophone. The limiting factor is cost. In order to realize this ideal, we have to make such communications affordable.

The developed nations will of course be the first to set up fiber-optics communications systems. It has been observed that the United States may lag behind in this development because of government antitrust regulations, which prevent any single corporation from controlling the market. However, this competition should result in a more cautious approach and ultimately superior system quality. Furthermore, some societies may be reluctant to get involved with a technology that makes information too accessible. But there will never be a shortage of the materials needed to make the glass for optical fibers. The lower cost of fiber optics compared with wire systems will encourage its use even in the less developed countries.

LASERS FOR DIRECT COMMUNICATIONS

Lasers suffer from the same limitations as other light and infrared sources for use in communications through the atmosphere, with the exception that the laser diverges very little or not at all. The more likely application of lasers for communication over long distances will be in space.

Radio and microwave signals can be readily beamed over long distance in space. However, the extremely complicated nature of computer-to-computer "conversations" at high speeds necessitates a very broad bandwidth that would be easily accommodated by a laser system. Moreover the laser is immune to interference or "jamming" unless an opaque object gets in the way of the beam. Communications signals might be impressed on a laser beam used for tracking or navigation. Only the desired signal destination would be receiving a given signal because of the narrowness of the beam; all other receivers would be locked out.

There is some evidence to suggest that lasers of certain wavelengths may propagate well for moderate distances under water. Presently, sound waves or very low-frequency radio waves are used for this purpose. Both of these emission types are incapable of carrying information at a rapid rate because of their low frequencies. Acoustic waves also suffer from the limitation imposed by their speed of propagation. The laser, while not providing exceptional range, does allow for high information transfer rate and is practically impossible to "jam."

For communications through the atmosphere, the helium-neon, laser diode and argon lasers are effective. The neodymium-YAG laser may also be employed. All of these systems operate at relatively low power levels, so that the eyes will not be injured if someone happens to look directly into the beam. Lasers are useful for communications between aircraft, which may have line-of-sight range in excess of 100 miles, and between ships and ship-to-shore. Lasers may also be used for satellite communications.

It is possible to communicate over paths not directly linked by line-of-sight using lasers. Reflectors of the first-surface type are used for this purpose. Such reflectors need not be large because of the small diameter of the laser beam; however, they must be oriented with precision. Even such passive objects as clouds may also be used to provide communications if the receiving apparatus is sensitive enough.

A possible method of using lasers for atmospheric-scatter or cloud-scatter communications is illustrated at FIG. 4-11. At the

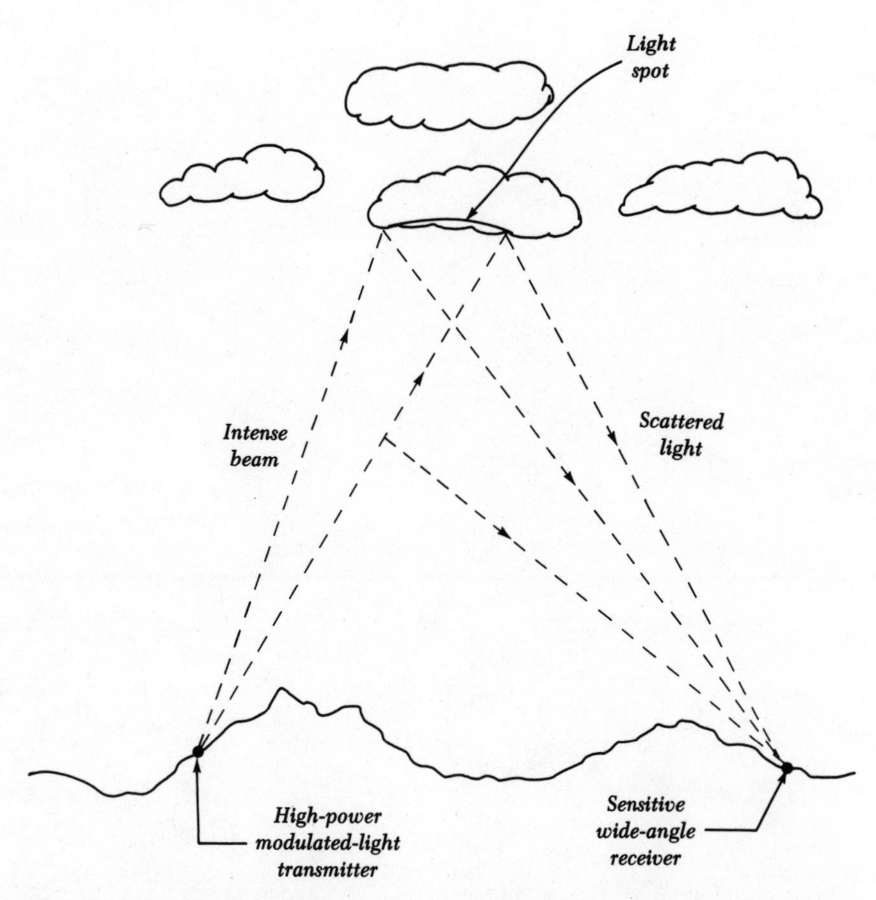

Fig. 4-11. Modulated light beams may be bounced off of clouds, or scattered by the atmosphere, for over-the-horizon communications using sensitive receivers.

transmitting site, a high-powered laser beam is aimed at the underside of a cloud bank or into the air. The receiver consists of a reflector of large diameter, a set of photocells or phototransistors, and high-gain, low-noise amplifiers. The receiver is aimed at the spot produced on the clouds or at some point near the horizon above the transmitter, where scattering by dust particles and water droplets would occur. The spot might not be visible to the unaided eye, but the large size of the receiving reflector and the high gain would retrieve the signals. In fact it would not necessarily require a laser source at all in order to achieve short-range communication by this means, provided the source could be sufficiently modulated at the desired rate.

5

Lasers in Industry and Entertainment

Lasers, because of their precise wavelength and because of the coherent light they produce, can be used in many ways to make exact measurements of distance, shape and time. The concentrated light with precise phase can produce bizarre visual images and patterns that make them useful for so-called "light shows" and other entertainment purposes.

In industry, lasers can be used for cutting, welding and drilling, soldering, component manufacture, and quality assurance. The concentrated energy, often thought of in terms of a "ray gun," can exceed the energy of the hottest blowtorches, and may completely vaporize one part of an object while having no effect on adjacent regions.

MEASURING DISTANCES

One of the obvious characteristics of the laser is that its beam travels in a straight line through space, barring refraction or reflection by intervening media. Ordinary tools for surveying and measurement fall short of this perfection. The laser can be used to construct perpendicular or straight lines, to bisect distances, or to geometrically construct anything that can be theoretically derived on paper by means of the mathmatician's traditional compass and straight edge. Examples are practically without limit.

The simplest way to measure a distance is with a pulsed laser, a mirror, and a timing device. This method does not necessarily require a laser, but the parallel rays of light from a laser travel great distances with very little attenuation, and this makes it possible to measure very large distances that cannot be accurately determined in any other way. The classic example is the measurement of the distance between the Earth and the Moon. These two planets are about 252,000 miles (400,000 kilometers) apart and light requires about 2.7 seconds to complete a round trip.

The apparatus for measuring distance with a laser is illustrated in FIG. 5-1. The pulse from the laser is first passed through a beam splitter, and a small percentage of the output is collected by a phototransistor amplifier. The electrical pulse is displayed on an oscilloscope or fed to an electronic interval timer. The greater part of the beam travels to the object in question, where a mirror or tricorner reflector is placed so that the beam returns exactly along the path it came. The orientation of the tricorner reflector is not critical, and will return any beam of light regardless of its arrival direction. The beam arrives back at the apparatus where a second beam splitter delivers it to the detection device. The distance D in miles is then determined by the equation

$$D_{miles} = 93141t,$$

where t is the round-trip time in seconds for any given pulse of the laser. The formula for kilometers is

$$D_{kilometers} = 149896t$$

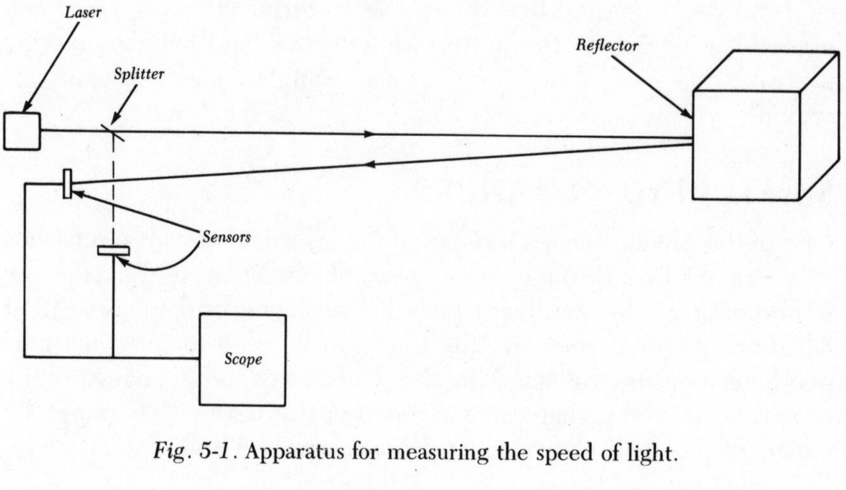

Fig. 5-1. Apparatus for measuring the speed of light.

The distance between the Earth and the Moon has been measured with such accuracy that changes can be detected because of the time of day or night, the presence of hills, and of course the irregularity of the Moon's orbit around the Earth.

Lasers can be used to measure distances between mountain peaks or between various points on the Earth. These distances vary because the land is not absolutely rigid. A mountain range may be rising by two inches a year because of the movement of a continent. This kind of change could not be detected accurately until the advent of light devices for distance measurement. While a change of just two inches a year may not seem like much, it amounts to 167 feet (50.8 meters) in 1000 years and 1,667 feet (508 meters) in 100 centuries. In geological terms, 100 centuries is a very short period of time.

Mountains rise and fall. Huge plates slide along each other. These movements are not usually smooth, but rather jerky, attended by violent events such as volcanoes and earthquakes. By observing patterns of movement of the Earth's crust in a particular place, and comparing these with the occurrence of violent upheavals, it may be possible to make good medium- and long-range predictions for major events. Lasers can determine changes in terrestrial displacements to a fraction of a millimeter.

The Laser Geodynamic Satellite, launched in 1976, gives exact measurements of these terrestrial distances. The satellite is covered with reflectors of the tricorner type which return light along the same line as it is beamed up. Laser pulses are directed at the satellite and their time of transit is measured from a specific point on the surface. The precise position of the satellite in the sky, as seen from the transmitting point, is also determined. This test is done from several locations around the world. From this data, the relative distance between any two points can be determined, by means of computer-performed calculations, to within a few centimeters. Changes can be detected as the tests are repeated at intervals. This reveals movements of the Earth's crust that would not be detectable by any other methods.

The correlation between movements of the earth and quakes and volcanoes is complicated, and scientists are only beginning to understand it. A major change in position may not mean a quake is imminent; in fact, the lack of significant movement for a period of time following uniform motion may be more indicative of a coming disaster, since plates in the earth tend to "catch" on each other and then suddenly "give."

SURVEYING AND LEVELING

Lasers can be of value in surveying, especially in the determination of straight lines. Light always follows the shortest path between two points as long as the index of refraction of the transmission medium does not change in the path, and in practice this is a straight line for purposes of construction. A laser can very accurately indicate the points along a straight line simply because the beam is always on

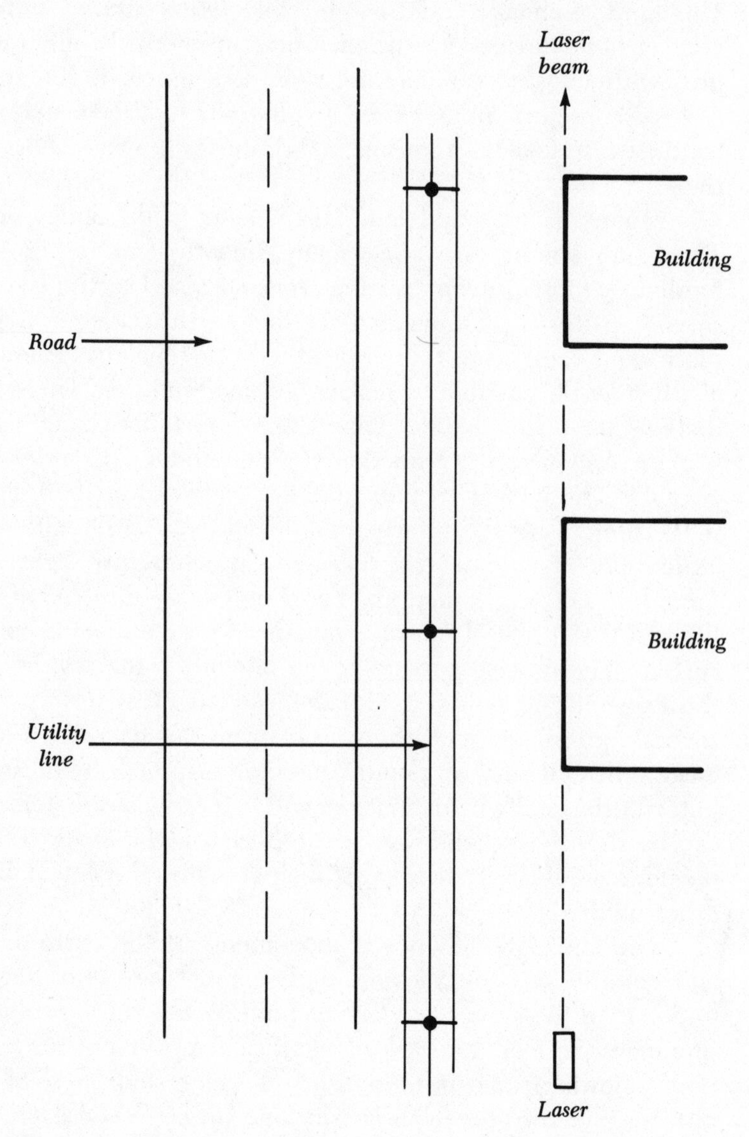

Fig. 5-2. A laser can be used to precisely align objects such as the foundations of buildings.

that line once the line has been set up by sighting between two end points (FIG. 5-2).

Lasers can be employed for determining where the foundation for a building should be laid, where a highway should go, where the property boundaries for a certain piece of real estate are, and other normal surveying applications, with considerably less time and effort then required by traditional surveying methods. Measurement of distances is important in surveying, and the laser provides accuracy that was impossible using other methods. Reference points in aerial photographs can be quickly and easily located using laser techniques for measuring distances and angles. This is useful in making localized maps of surface features and topography with accuracy down to a few millimeters—the limit of resolution for paper maps, and making it conceivable to use computers for mapping large areas with microscopic resolution.

MEASURING SMALL DISTANCES

Lasers are also used for very precise measurement of small distances, since they emit discrete wavelengths of energy on the order of nanometers (1 nm = 10^{-9} m). The technique used for measuring microscopic displacements is called interferometry and the device employed is called a laser interferometer.

Waves have the property of producing regular patterns as a result of phase addition and cancellation. Perhaps you have seen this effect in a ripple tank, or you are familiar with the directional properties of radio antennas, or have worked with optical diffraction gratings. The particular pattern produced depends on the wavelength of energy and the separation between sources or reflected beams. Since light has such a small wavelength—red light, for example, is represented by about 7×10^{-7} meter or 700 nm—tiny displacements will result in inteference patterns. The number of bright and dark spots or bands in an interference pattern indicates how far apart the sources of light are.

Waves that have equal amplitude and equal wavelength may add together in phase (FIG. 5-3A) or completely cancel each other (FIG. 5-3B). If they arrive in some intermediate phase relationship, then the amplitude of the result will be somewhere in between that of complete addition and complete cancellation. Phase differences may be measured in fractions of a wavelength, degrees of phase, where zero degrees is in phase and 180 degrees is out of phase, or in radians of phase.

The output of a laser is such that all the waves are of the same

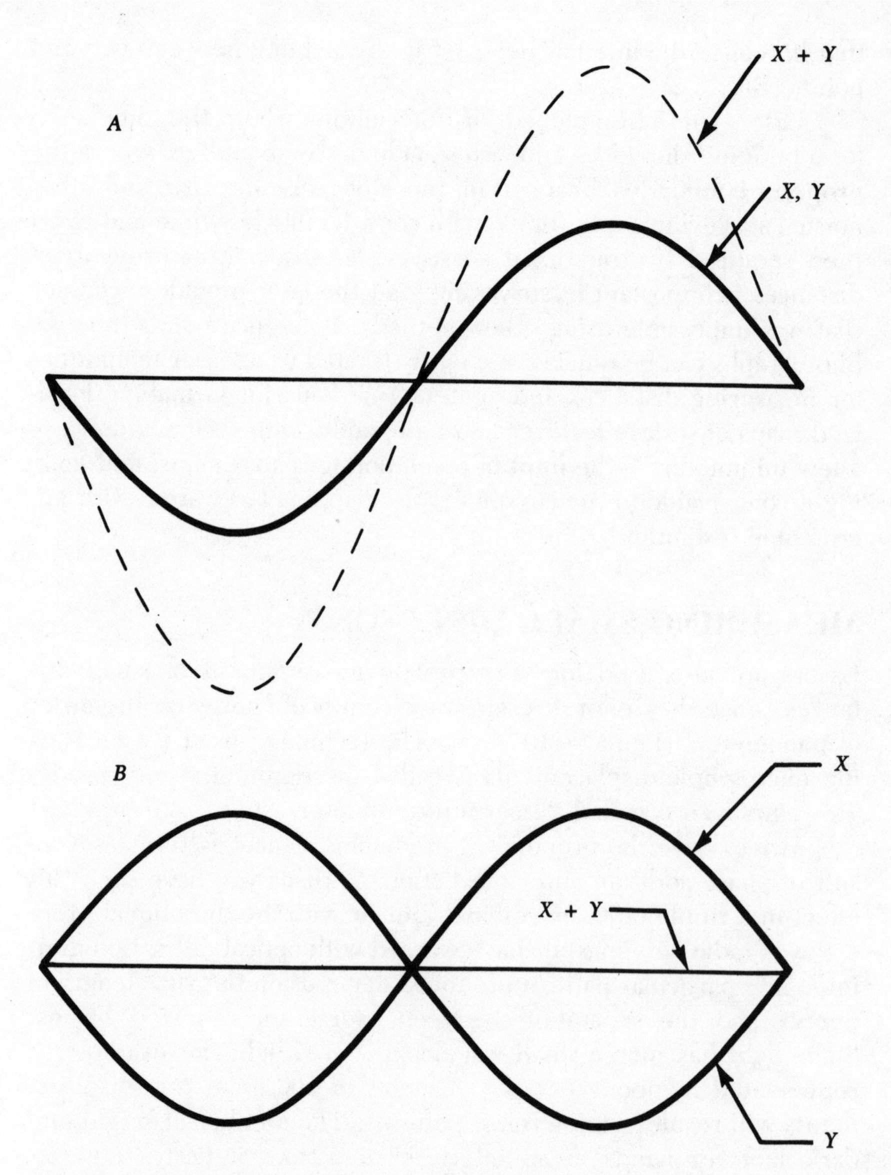

Fig. 5-3. (A) Two waves of equal amplitude and frequency add in phase to produce a sum having twice the amplitude of either. (B) When the waves are in opposing phase they cancel completely.

wavelength and are in phase (coherent). Therefore the laser is especially good for producing interference patterns. If a laser travels over two different paths to the same target, and the paths differ in length, there will be interference at the target. Usually this will take the form of a pattern of alternating bright and dark bands. The interferometer consists of a laser light source, two mirrors, a beam splitter,

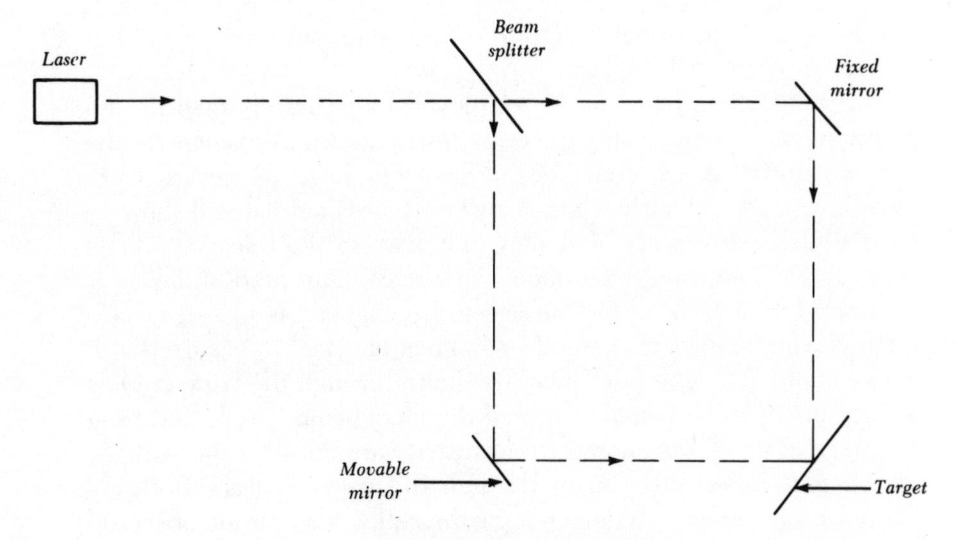

Fig. 5-4. An interferometer.

and a target. The beam splitter transmits half of the laser energy and reflects the other half at some angle. Each of the beams is aimed at one of the mirrors, and this mirror reflects the beams to the target where an interference pattern is produced. The other mirror is moved while one is left fixed (FIG. 5-4). The movable mirror is displaced slowly over the distance to be measured, and as this is done, the light at the target gets brighter and darker at one particular point. The number of alternations is counted. Each cycle from light to light or dark to dark indicates that the path difference for the two beams has changed by one wavelength. Knowing exactly the wavelength of the light used and the particular geometry of the interferometer arrangement, the displacement of the movable mirror can be determined with great precision.

Interferometers are more often used to ensure that, in practice, given geometrical dimensions of apparatus stay the same. Temperature changes, mechanical vibration or blows, and normal use can all cause the alignment of equipment to become inexact. Interferometers show clearly when a dimension is not correct, since the appearance of the pattern will change dramatically if movement occurs even over a part of a single wavelength of the laser light. This is especially important in a machine shop, where interferometers are set up at the beginning of the day or at periodic intervals, and the intensity of the spot or nature of the pattern is frequently observed. The interferometer may also be used in conjunction with computerized robots that automatically compensate for changes in dimensions and bring the equipment back into proper alignment before any significant deviation takes place.

The interferometer can be employed to accurately map the topography of a surface. Tiny irregularities show up as changes in the fringe pattern. A perfectly flat surface will have fringes that are equally spaced and fairly wide. A surface that is not flat will show irregularly shaped fringes, and they may vary in their spacing, some even perhaps appearing as concentric circles. This kind of device is illustrated in FIG. 5-5. The surface to be checked is placed underneath a pane of glass that has already been checked to ensure that it is essentially flat. The laser beam is shone through the pane of glass at the surface to be tested. Some of the laser beam is reflected from the pane of glass, but most is transmitted through it to the surface, where it is reflected back up through the pane of glass. Both reflected beams meet at a detector, or the reflections can be observed

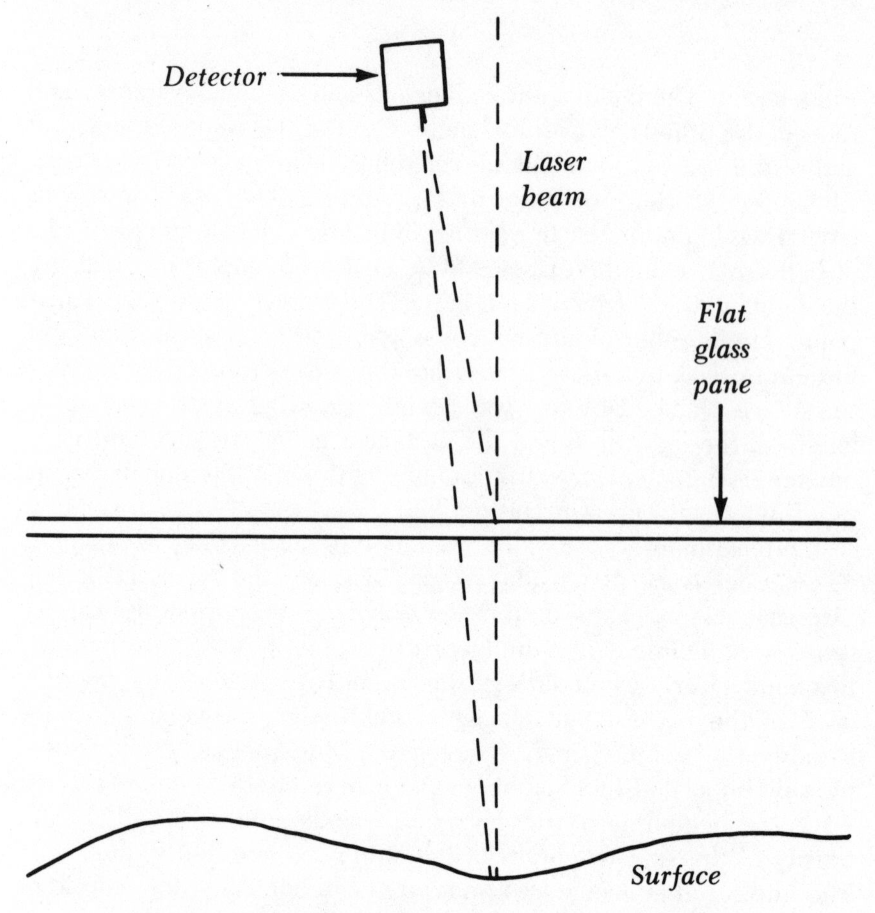

Fig. 5-5. Apparatus for determining the flatness of a surface.

as they fall on a viewing screen. A perfectly flat surface is indicated by widely spaced fringes having regular contour. Bumps or hollows are indicated by irregularities. The technique is similar to what occurs when you look through two screens. Irregularities in this case are intuitively obvious even when the observer is not familiar with the principle involved.

Interferometers are frequently used to check the lenses and mirrors in telescopes, cameras, microscopes and other optical equipment. Astigmatism, or the tendency of a lens to have a different focal length for light in one plane of polarization as compared with another plane, is easily detected. A standard sample, known to be essentially perfect, is compared with the specimen under test. The two are placed close together and the light transmitted and reflected interferes in the same way as in the example of Fig. 5-5. Deviations from perfection are easily seen as irregularities in the pattern of interference. Incoherent light was once used to test optical devices in this way, but helium-neon lasers are generally used today since they produce better interference patterns. The smallest deviations can be detected and astigmatism practically eliminated in mirror-grinding and lens-grinding processes.

ENSURING SMOOTH EDGES

Lasers are also employed for making sure that edges are smooth and of the correct shape. The points of surgical needles and the edges of knives, for example, must be smooth and free of rough spots.

Rough lens edges can be detected in either of two ways. One method uses a small beam directed at a spot on the lens edge: rough spots cause irregular reflection of the light. Another method employs two laser beams that intersect at the point where they strike the lens edge: irregularities in the interference pattern indicate rough spots. The same techniques can be used for knife edges or other edges that require precision smoothness. For hypodermic needles, samples are compared with a standard needle known to be perfect. The light from a helium-neon laser is directed at the needle tip, and the tip scatters the light in a certain way. The needles are inspected at high speed by means of sensors and a small computer. Needles can be sorted very quickly and efficiently in this way.

EXAMINING SOLDER JOINTS

The laser provides a way to actively test the behavior of solder joints under thermal stress. In this method of checking, the laser actually heats the material to be tested, and the cooling curve is plotted and analyzed using a computer. The system was developed by Vanzetti Infrared and Computer Systems, Inc.

A neodymium-YAG pulsed laser is employed. The pulses of laser emissions heat the solder joint but do not actually liquefy the solder. So the joint is heated but not disturbed. The behavior of cooling is monitored using an infrared detector. A curve results, and this curve can be compared with known curves for normal joints and for joints with specific kinds of defects. If there is a problem with a particular solder joint, such as its being "cold" (not enough heat used when the joint was made), the defect can be found quickly and its nature can be determined precisely.

EXAMINING CLOTH

In the textile industry, fabrics have traditionally been inspected manually. The work is tedious and quite time-consuming, and therefore expensive. Lasers are providing a way to check cloth for defects much more quickly than has been possible in the past. In some cases the speed is up to 55,000 miles (88,000 kilometers) per hour.

The Ford Aerospace and Communications Corporation was one of the first companies to use lasers for scanning cloth. Helium-neon lasers were employed. The laser energy is split into three beams. Each beam scans at about 55,000 miles (88,000 kilometers) per hour. The fabric moves by at a speed of about 12 feet (4 meters) per second as the laser beams scan across it. If a flaw is found, ink is squirted on it from a high-speed nozzle. The laser makes it possible to determine the nature of flaws as well as their mere presence and location. All this is recorded by a computer for future reference. If the defect can be repaired without cutting out a piece of cloth and replacing it, this is done and the laser device indicates that it is possible. This saves more time and work.

Lasers are widely used in examining materials along assembly lines. All of the objects on a certain conveyor belt are supposed to be identical. This can quickly be verified by shining laser beams at each object and recording the scattered light. An example might be measuring the height of jars. A laser is shone just over the tops of the jars as they go by. Another laser is shone just below the level of the jar tops. The first laser beam should never be broken, while the second beam always should be. Any deviation from this pattern indicates a defective jar. The jar is found and removed by a robot.

GUIDING SAWS

In the lumber industry, laser beams are used to show the worker exactly where a saw is going to cut a log. This makes it easy for the worker to guide logs through cutting machines. This device is very inexpensive and there are thousands of them currently being used in the lumber industry.

The laser draws the line by means of a fast scanning mirror (FIG. 5-6). The beam from a helium-neon laser is shone at the mirror, which rotates very rapidly so that the beam describes a plane in space, corresponding to the plane of the cutting saw. The beam scans several times a second so that the light appears continuous and looks like a line instead of a fast-moving spot.

Laser scanners can be used to precisely measure the dimensions of logs. This information is then fed to a computer, which determines the best way to cut the logs in order to get the most possible lumber from them. This technique is especially useful in the cutting apparatus used in making plywood. Sheets of about ⅛ inch

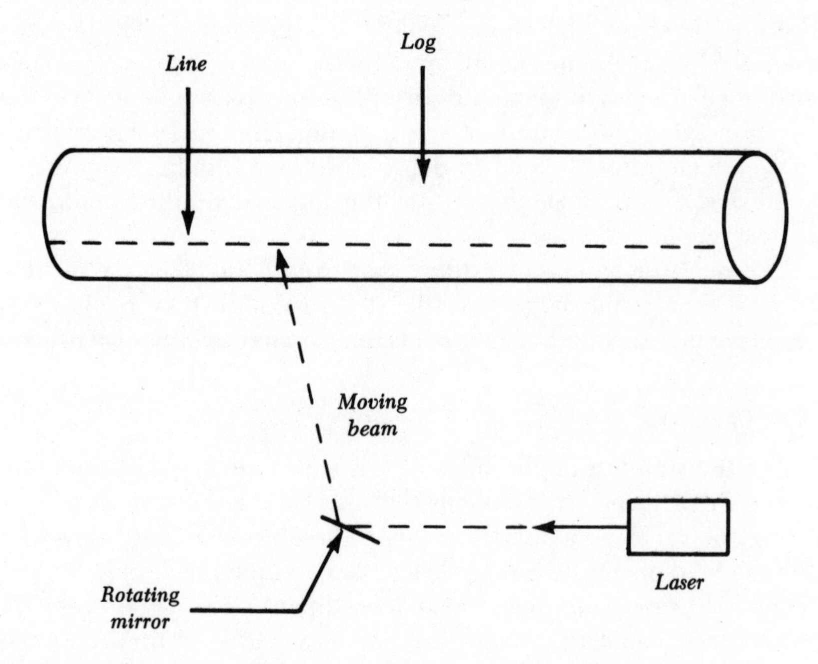

Fig. 5-6. A laser can be used to "draw" a line on a log, indicating the path through which a saw will cut.

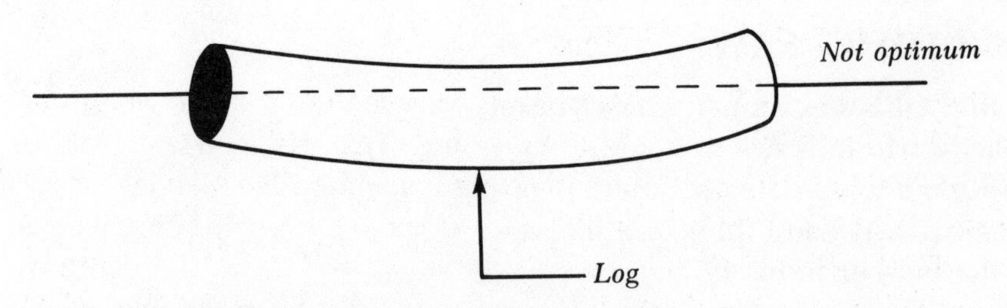

Fig. 5-7. The straight line connecting the centers of the ends of a log may not be the optimum axis for making plywood veneer.

(3 millimeters) in thickness are peeled from logs by rotating the logs in a special device with a long cutting blade. Accuracy of log placement is critical here, since small errors may result in uneven thickness of the veneer and consequently in defective plywood or even useless veneer.

Laser scanning of logs increases the amount of veneer obtainable by as much as 10 percent over older methods. In the traditional method, a log was rotated about its two ends and the straight line described between the chosen points was not necessarily the axis that would yield the most veneer. This was simply because the logs were not perfect cylinders and an optimum axis might not be the one connecting the centers of the log ends (FIG. 5-7). Laser scanning and computers helped to determine this optimum axis, which would describe the cylinder within the log that would contain the greatest amount of wood.

These laser scanning systems cost from $100,000 to $250,000; but in the long run, large sawmills stand to more than recoup this cost in savings of wood, which is becoming increasingly expensive.

COUNTING

One of the simplest applications of lasers in industry is in counting items at high speed. It is obvious that this can be done on an assembly line by breaking a laser beam with the objects as they move by. This can be done at the rate of dozens or even hundreds of items per second. The laser counting system is used in the newspaper industry to accurately count the number of papers printed each day. As the cost of these counting systems decreases, it is expected that their use will be commonplace even in smaller newspaper publishing establishments.

MEASURING ROTATION

Lasers can be used for other measurements besides those involving straight lines or numerical counting. Gyroscopes can employ lasers for increased accuracy. These devices are used in aircraft and guided missiles for determining course.

The laser gyroscope consists of a resonator containing three mirrors, arranged at the vertices of a triangle, rather than the usual two mirrors in the conventional laser tube. The mirrors are aligned perfectly so that the light beams go exactly along the sides of the triangle. Two laser beams are used, one in each direction (FIG. 5-8).

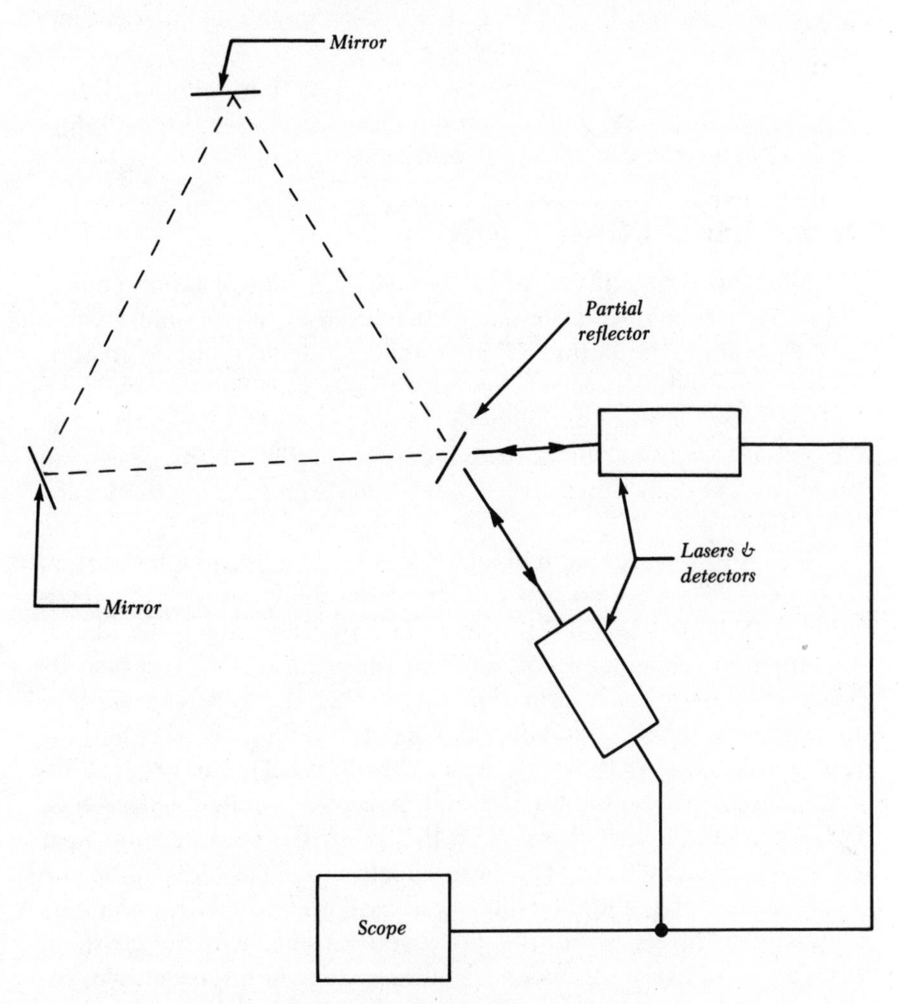

Fig. 5-8. Apparatus for determining rotational motion.

Rotation can be sensed around the closed path because of slight differences in the behavior of the two laser beams. This is because, as the system rotates, the mirrors move a little with respect to the light beams, which always must describe straight lines through space.* As the two beams go around the circuit, one beam must go farther than the other between each mirror when the system rotates. This difference can be measured even for very slow rates of rotation if the measuring apparatus is large enough. This speed difference shows up as a phase difference, since the waves traveling the greater distance require more wavelengths to cover it than the waves traveling the lesser distance. The single-wavelength laser emissions make possible very accurate determinations of rotational speed.

When three laser gyroscopes are used together, complex three-dimensional rotational motion can be observed. These three components in an aircraft are tilt, pitch and yaw.

DETECTING POLLUTION

Pollution and contaminants are detected by means of laser spectroscopy. The technique is similar to that used by astronomers when they determine the nature of interstellar and intergalactic matter. Various materials absorb energy at specific wavelengths while allowing energy at other wavelengths to pass virtually unaffected. The absorption spectrum, or pattern of wavelengths in the spectrum caused by a certain substance, is unique and can be used to identify that substance.

The output wavelength of a laser can be tuned through a certain range of wavelengths, so that the absorption lines can be very accurately determined and the presence of unwanted chemicals identified in certain gases and liquids. The presence of ions can also be detected. This gives information concerning the behavior of substances under different conditions. Changes in temperature, and exposure to ionizing radiation such as solar ultraviolet, can produce increased concentrations of some substances. A familiar example of this is the production of ozone in the air in the summertime near large metropolitan areas. The laser spectroscope can determine the concentration of pollutants such as these, and when the amounts reach certain levels, advisories are given. Ozone, which consists of three atoms of oxygen grouped together instead of the usual two, re-

---

*This is based on space being Euclidean, neglecting the relativistic effects of gravitation or acceleration.

sults from ultraviolet action on automobile exhaust in the air. This is ironically the same chemical, the depletion of which is becoming a major concern to environmentalists. Ozone irritates the bronchi in suspectible persons, and can cause difficulty in breathing. The chemical has a smell something like that of chlorine bleach.

The laser spectroscope has high resolution—much higher than traditional spectroscopes can provide. This makes it possible to detect such phenomena as the splitting of spectral lines by a stable magnetic field.

Other pollutants, such as sulfur dioxide, can be detected and the concentrations measured. Sulfur dioxide is a byproduct of the combustion of coal, used in some electric power plants. Lasers can be used to detect contaminants in drinking water as well.

Another method by which lasers can detect pollution is by measuring the scatter caused by particulate matter. A laser normally produces a thin, straight beam of light that does not diverge very much. Dust and other particules in the air cause scattering and may also cause greater attenuation with distance. These phenomena can be measured and can give a clue as to the quality of the air in a given sample.

Lasers have the advantage of being extremely sensitive for monitoring contaminants and pollution, at times giving indications down to a few parts per billion. Under laboratory conditions, even a single atom has been detected using a laser.

Lasers are not yet in widespread use for measuring air pollution, although in larger cities they have been used. More often, laboratory experimentation, such as the testing of pollution-control devices for automobiles, has been the chief role of lasers in pollution research.

DRILLING HOLES

The laser will probably always carry with it the connotation of a boring or cutting device, capable of penetrating solid metal or concrete. This is one useful property of high-energy lasers, and the devices have been used to drill precise, sometimes very tiny holes in manufacturing processes.

The first holes were drilled using a ruby laser. The beam was focused to a sharp, tiny point, concentrating all the energy into a minute area. A convex lens was adequate for this (FIG. 5-9). The principle was exactly the same as that used in solar cookers, except that much smaller lenses could be used since the laser beam was almost perfectly parallel, mimicking a point source at infinite distance.

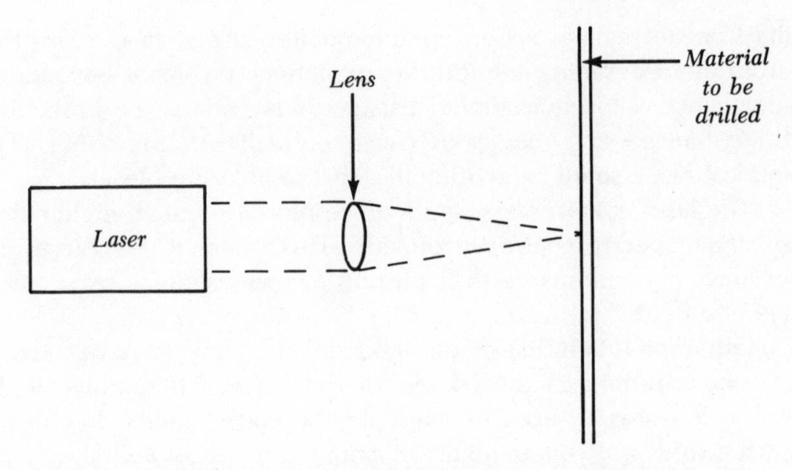

Fig. 5-9. A lens is used to focus laser energy for drilling holes.

This is one of the chief problems in getting a large amount of energy from conventional light sources: the energy is difficult to concentrate. Not so with the laser.

The immediate impression was that lasers were highly powerful devices, capable of burning, vaporizing or totally destroying things from vast distances. Immediately, these "ray guns" became popular topics for science fiction. Laser technology allowed more convincing descriptions and illustrations of these weapons. All one had to do was to take the laser with its helical flash tube and ruby rod, and make it as big as a house, and—boom!—down would come aircraft, and huge warships would be split apart amidships and go bubbling to the bottom. The reality was too tame, too disappointing.

One of the main advantages to using lasers for hole-drilling is that the drill does not touch the material being drilled. There is no drill bit to break, slip, clean or replace. Tiny, or even microscopic holes can be made without worrying about breaking a thin, weak drill bit. Brittle, fragile or extremely hard materials can be drilled easily without damage to either the drill or the material being drilled. Soft materials like rubber, which can be moved out of place when mechanical drills are used, resulting in a hole that is irregular, are handled more readily since the laser does not involve physical contact. There is no friction with the laser; heating of surrounding material is kept to a minimum. Places that are difficult to reach with ordinary drills are often easily accessible using the laser.

These advantages are offset by the high cost of a laser drilling apparatus. A pulsed laser and support equipment costs in the tens of thousands of dollars. Personnel have to be trained to use the laser drill. Also lasers are not very good at making larger holes, since the amount of energy needed increases roughly according to the area of the hole, and this is proportional to the square of the diameter. For example, in a given material, the energy required to drill a 5-mm hole is 25 times that needed for a 1-mm hole.

New technologies are sometimes slow to be accepted. The worker often prefers to stay with the "tried and true" methods, not wanting to have to learn new things. The manager is concerned, too, about the cost of training in terms of time and lost production. The laser is a delicate instrument, too, and must be maintained and protected from damage from dust and corrosion.

LASERS IN MANUFACTURING

There are various applications of the laser in mass manufacturing. We can categorize the uses of the laser according to the amount of energy needed, such as high, medium and low. The greatest energy is needed for such tasks as drilling holes and cutting through large objects. Much energy is also needed for welding. At the low-energy extreme are the interferometer and counting devices. In the middle range are applications such as writing serial numbers on production units, small-scale soldering, and communication devices such as optical-fiber networks that might be used to interconnect computers in a factory.

Lasers in manufacturing may be either continuous or pulsed. Continuous lasers may be modulated if they are used to convey information. Drilling is done most often using a pulsed laser, while cutting, welding, soldering and communicating are done with continuous lasers, although the pulsed laser can be used for almost any industrial application. Annealing is generally done with a laser having a continuous output.

The types of lasers most often used in manufacturing are the neodymium-YAG, ruby, and carbon-dioxide devices. Neodymium-glass lasers may also be used. These are all high in energy output. The carbon-dioxide laser produces the greatest energy in a continuous beam; the others are pulsed.

In the manufacture of wire, holes are drilled in diamonds and the wire is then pulled through the holes. The holes themselves are drilled by means of lasers, and a single hole requires only a few minutes to produce. This compares with several hours or even days before lasers were used. The hardness of the diamond is such that it is very difficult to scratch, cut or drill by mechanical means, but a high-energy laser of the correct wavelength will simply vaporize it. Mechanical methods are still used to polish the hole after it has been drilled since the laser usually leaves a rough or irregular hole.

Lasers are also used for drilling and cutting ceramic materials, which are widely used in electronics because of their dielectric properties and their physical ruggedness. The problem with mechanical tools when used with ceramics is not a matter of hardness, but of fragility: ceramics break easily. Since there is no mechanical stress induced by the use of lasers for cutting and drilling, but only some thermal stress, there is much less chance of a mishap when lasers are used in place of older methods for ceramics work.

Lasers are used to drill holes in circuit boards, and this has been a breakthrough in mass manufacture of printed circuits. The glass-epoxy type of board, especially, lends itself to laser drilling since the material is fragile. In mass production, laser devices are programmed and operated by robots so the holes are automatically drilled in the right places at the correct size, over and over. Lasers are also used to make perforations in materials so that they can be broken apart without shattering or other damage. The holes are drilled along the line where the material is to be broken, and penetrate only part of the way through.

Lasers are used to drill holes in softer materials, too, such as rubber. Such materials will often yield to the frictional pressure of mechanical drilling or cutting, and this will result in irregular holes or cuts and may cause such problems that it becomes impossible to use mechanical means on the material. You may have encountered this problem when trying to cut rubber with a scissors, for example. Large holes are especially well suited for laser drilling when the material is soft.

Carbon-dioxide lasers are widely used for drilling in rubber, and were first introduced over twenty years ago. The laser does not have to be especially powerful; often a few watts is all that is needed. Lasers are used to drill holes in such things as aerosol valves and baby bottle nipples.

Very tiny holes present a problem for laser drilling in soft materials, so these materials are often made with pins in them where the holes are intended to be. However, the pins tended to break when they were very tiny indeed. Since laser technology has been perfected, the smallest holes can be made at the rate of more than 10 per second in actual production, using lasers of about 150 watts.

Small holes are punched in a variety of other substances, too, using lasers. A good example is cigarette paper. The holes are drilled to regulate the air flow that the smoker gets, thereby minimizing the tar content in inhaled smoke. The holes are about 0.2 millimeters in diameter, so small that a magnifying glass or low-powered microscope is needed to see them. The holes can be punched at the rate of hundreds of thousands, or even millions, per second using a carbon-dioxide laser. Lasers are used to drill holes in capsules for the timed release of certain pharmaceuticals. The holes must be of just the right size so the medication is absorbed at the proper rate. Lasers provide a predictable hole size, fast production, and low failure rate, which is crucial when drugs are involved.

Laser drilling has many advantages, mainly rapidity of production, exact predictability and repeatability. But there are also certain disadvantages to laser drilling. Metals are hard to drill because they reflect laser light. Glass is also difficult to drill since it tends to transmit the laser light instead of absorbing it. This makes it necessary to use high-powered lasers to ensure that enough of the material absorbs the light. A substance that reflects or transmits 95 percent of the laser light, and which will melt when it gets 50 watts of laser energy, will require a laser of 1000 watts to be melted.

Material that is melted, or even vaporized by a laser does not disappear but will reform near the hole. It may condense on anything near the hole, whether it be the substance being drilled, or something else. The material around the hole may be partially melted and may not reharden evenly, causing an uneven or irregular hole.

Probably the most significant problem in laser drilling is that, since the laser beam must be focused by means of a lens, the hole does not have perfectly parallel sides. The hole tends to resemble a cone more than the long, thin cylinder we would ideally get. This problem can be made less severe by using a lens with a longer focal length, as shown in FIG. 5-10. The shorter focal length, shown at A, results in a cone with a large apex angle. The longer focal length, at B, reduces the apex angle and the hole more closely resembles the ideal.

Lasers cut in the same way that they drill, that is, by melting or more often by vaporizing the material to be cut. The vaporized material must be blown away by a gas jet so that it will not recondense and cause problems. Materials that reflect or transmit the laser light are burned away by using a jet of oxygen at high speed. This encourages oxidation of the material and also helps to drive away the debris so that the cut will be clean. The laser does not actually do the cutting in this situation, but serves simply to heat the material to a temperature sufficient for burning when the oxygen is supplied. The process is similar to that of fanning hot coals into flame. Lasers are

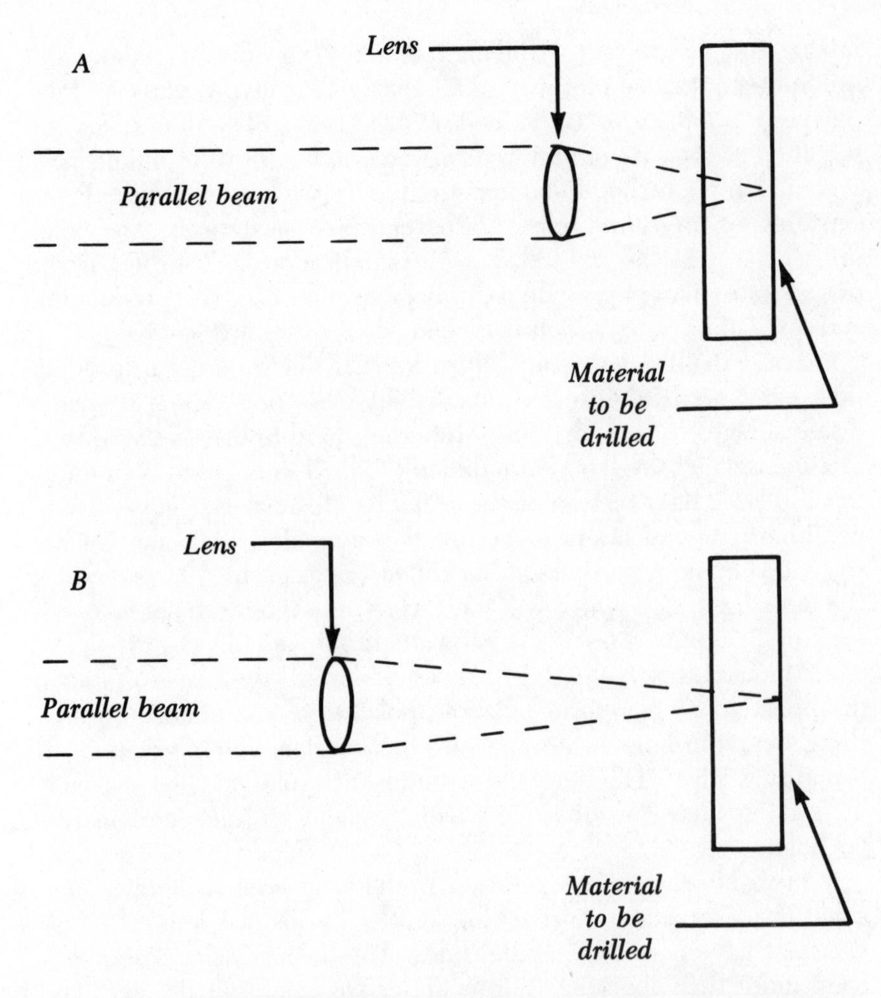

Fig. 5-10. (A) A lens with short focal length produces poor depth of the hole. At B, the hole has greater depth because of the longer focal length of the lens.

used to cut various metals, especially titanium, since this metal is difficult to cut by other means because it has a low igniting temperature.

CUTTING AND WELDING

A laser can be used to cut just about anything that will absorb any radiant energy at all, which includes wood, paper, plastics, rubber and even cloth. In some cases the older methods of cutting have proven more effective and cheaper than the laser which, although a "sexy" tool for cutting, may be akin to shooting a sparrow with a cannon.

In the case of wood, the laser didn't work out very well as a cutting tool because it could not cut very deep and was expensive. The heat from the laser causes the wood to be charred near the cutting edges, a problem that generally does not occur with traditional metal blades unless they are operated at extreme speeds with hard wood. In some cases, where pieces must be cut to exact specifications, the laser is used, but we will probably never see lumberjacks cutting down trees like Darth Vader wielding a powerful laser gun.

Lasers are used to cut flexible plastics and rubber that is deformed easily by mechanical tools. The problem is melting, and the resulting irregularity of the cut when the material rehardens. Lasers can seal as well as cut certain kinds of plastics if the correct temperature and rate of cutting are used.

Lasers can cut patterns from pieces of cloth very rapidly and accurately, and laser machines are being used in the textile industry to perform tedious work previously done by people. The main difficulty, as with laser wood cutting, is burning or charring of the material. Some fabrics, such as synthetics, tend to melt instead of burn, and this can be a problem. However, if the cutting is done rapidly and air blowing or suction devices are used to speed cooling and remove smoke, charring and melting can be avoided in most cases.

Lasers can easily be used for marking purposes. They can produce an indelible pattern, or inscribe a serial number or identification code on an object. Extremely hard materials, such as glass, quartz and some metals, lend themselves to this process since mechanical inscription is particularly difficult for these substances. The process is much quicker than traditional methods and may be less expensive when done in large quantities. Automation reduces the cost still further. The writing may be done either with a continuous or pulsed laser. The pulsed laser leaves a string of dots, so that the resulting characters look as if they were made by a dot-matrix printer (FIG. 5-11). The dots may form recognizable characters or be arranged in a sequence that is encoded to represent something. Continuous lasers may be used for etching or engraving. Sometimes the entire image is etched at once, by passing the laser energy through a stencil.

Lasers are sometimes used for stripping wires or cables. Coaxial cable lends itself nicely to laser stripping. Coaxial cable has a center conductor surrounded by a cylindrical wire braid, with insulation (the dielectric) between the center conductor and shield, and a jacket outside the shield. If you have worked with this kind of cable you know that it is not especially easy to strip. Lasers make short work of it for such purposes as connector installation in mass quanti-

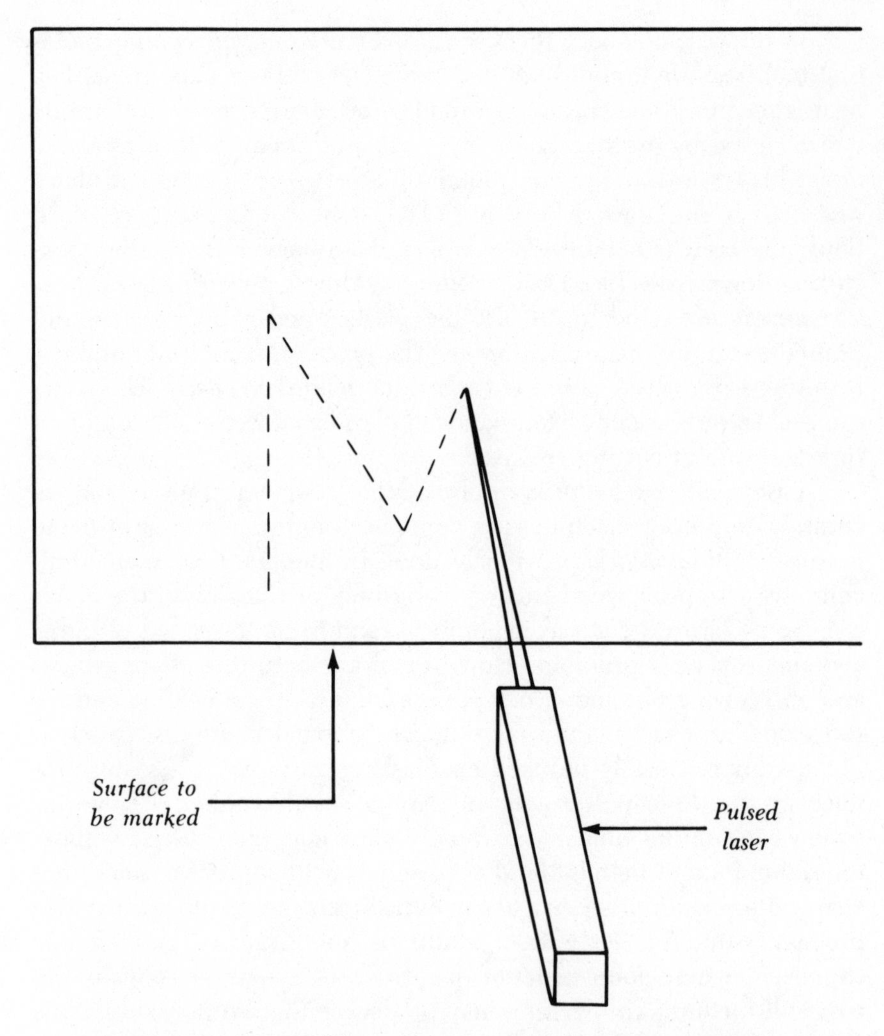

Fig. 5-11. A pulsed laser produces a dotted line when etching.

ties. Mechanical stripping is likely to nick the braid. The laser does it without damaging the braid or center conductor. The plastic or polyethylene insulating substances absorb the laser energy readily, while the metal conductors, usually copper, do not. This is very useful when small coaxial cables are used in cramped circuits. Such cables are easily damaged and are very difficult to work with mechanically.

Lasers can be used to cut and drill metals, as we have seen, although the energy requirements are high because the metals tend to reflect rather than absorb most of the energy. Welding is more practical; the energy requirements are less since the metal need only be melted, and welding metals is often done with substances that have fairly low melting points. Welding can be done in very cramped situations when a laser is used. It can also be done in a vacuum. When welding is done on an extremely small scale, the process is called microwelding.

If the welded material is provided with a heat sink, small amounts of metal can be made to melt without vaporizing as they normally would. Sometimes inert gases such as argon are used in the welding process to prevent oxidation (burning). The laser can be focused on a microscopic point, making it feasible to work with microchips and perhaps even repair some kinds of damage to such integrated circuits. Space and military programs are especially interested in laser welding of electronic circuits.

In the manufacture of transistors, different kinds of metals or alloys must often be welded at junctions far tinier than the point of an ordinary pencil. The laser does this with a minimum number of accidents, improving the quality and reducing the rejection rate. This is economically beneficial. Some laser welding devices are still in the experimental stage.

In large-scale welding applications, the traditional blowtorch and arc are more commonly used, as it is too difficult to obtain the necessary power to make lasers practical. This, too, may someday change, especially for welding in outer space, as in the construction of space stations. The main problem with high-power lasers is that they require bulky power supplies. This problem is difficult to overcome by means of solar power, although nuclear reactors could provide the answer. Nuclear waste disposal would not be a particular problem in interplanetary space.

Pulsed lasers are more commonly used in welding than continuous lasers. Either type will work theoretically, but high energy is easily obtained and regulated with the pulsed laser. Pulse rates range from about one every 10 or 12 seconds to once per second. The rate and pulse duration, as well as the pulse energy, determine how much the laser energy will be (FIG. 5-12). The higher pulse rates and larger pulse widths produce more power in a given system than smaller widths or shorter duration, even with the same peak energy.

Lasers can be used for soldering as well as for welding. Soldering is, in fact, a low-temperature form of welding using tin-lead alloys that adhere readily to some metals, especially copper. Soldering of printed-circuit boards can be tedious and difficult when done manually, but laser robots can carry out this task in mass production, with double-sided as well as single-sided circuit boards. Lasers will never replace the home workbench soldering gun, but they are

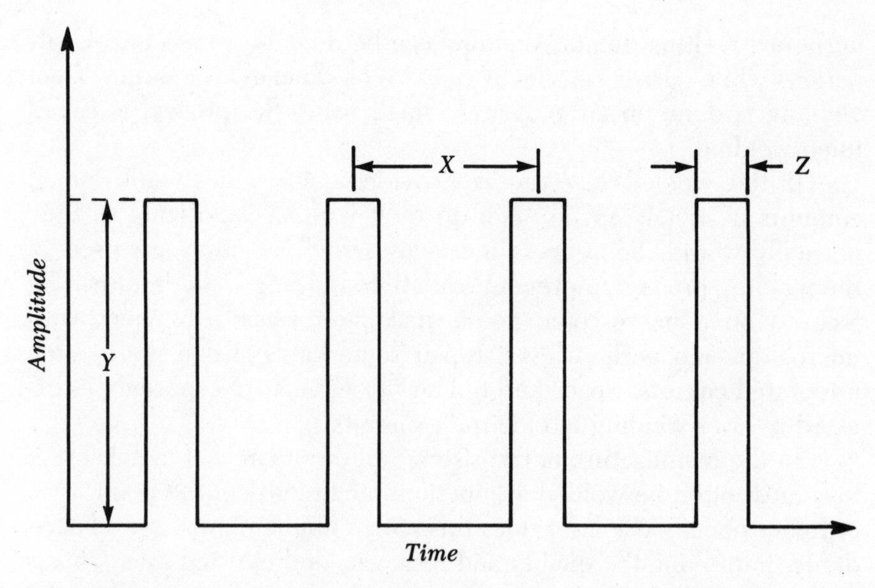

Fig. 5-12. Pulse frequency (X), amplitude (Y) and duration (Z) all affect the average energy output of a pulsed laser.

already being used in industry. Small, pulsed carbon-dioxide lasers are the most commonly used for soldering. Laser soldering is faster and more precise, and can apply exactly the same amount of energy to every joint. The amount of applied energy can be programmed into a computer. A laser can remove wire insulation at the same time that it heats a joint for soldering. "Cold" or overheated joints are essentially eliminated.

Lasers are used in the annealing, or heat treating, of metals for wire. You have probably noticed that wire is available in a variety of different hardnesses. Soft-drawn wire is easy to work with since it holds its shape when bent. But this kind of wire has relatively poor tensile strength and will stretch to some extent, a poor characteristic for use in antenna systems. The same wire that is ideal in one application may be very poorly suited for another. Thus we have hard-drawn copper wire for use in such things as antenna systems where tensile strength is important. In order to obtain a certain hardness, wire is heat treated or mixed with various other metals to form alloys.

Heat treating of metals often changes the way the atoms in the substance are arranged. A vivid example in nature, although not a metal, is provided by the element carbon. When the atoms are arranged one way, we have the black substance called graphite, used for marking and for lubrication. When the atoms are arranged another way, the same substance becomes diamond. Metals can be made harder by heating or by application of tremendous pressure. Generally, heat is easier to apply and is therefore used more often, especially in annealing wire. Continuous carbon-dioxide lasers provide one way to heat-treat metals.

In order to enhance the absorption of the laser energy, the metal to be treated is coated with a substance that is a good absorber of energy at the wavelength of the laser. The more heat that is applied, the deeper the heat treating. The amount of heat depends on the energy of the laser, the scanning rate, and the extent to which the absorbing substance gets heated. Laser annealing is energy efficient and can be done in areas that are difficult or impossible to reach by older methods.

THINGS LASERS CANNOT DO

There are certain things that we will probably never see a laser doing, simply because there are some applications to which the laser is not well suited. Cutting down trees is one example. Mowing grass is another rather silly example of something that won't be commonplace (can you see the sign: "Laser Lawn Mowers $2,995.00"?). Cutting food is something that has been tried and found difficult or impractical. It generally burns the food and costs more than it saves. Mining with lasers has been found impractical as well. It might at first seem like a very nice idea, especially for use on other planets. Maybe in some cases this will prove useful, but explosives are, for the time being, cheaper and quicker than any other known means. Even though the vision of a laser, being fired from a spacecraft at some barren, remote alien mountainside, complete with inaudible explosion (no air to carry sound) against black sky (no atmosphere to scatter starlight) is quite fascinating, it's more than likely that such explosions will result from plain old gunpowder.

LASER LIGHT SHOWS

Planetariums frequently give shows using lasers, with accompanying music, to dazzle audiences. Sometimes ordinary lights can be used and lasers aren't really necessary, but in some shows, three-dimensional images are formed in midair by techniques similar to holography. Lasers are required for this, because it is the interference between different beams and the narrowness of the beams that produces the images.

Laser pictures are generally drawn by tracing the outline of a geometric figure, the laser moving very rapidly by means of mirrors that can rotate in two dimensions. Such a mirror can be mechanically moved much faster than could be possible if it were necessary to rotate the whole laser apparatus (FIG. 5-13). Prisms may be used to split the laser into different color constituents when a krypton laser is used. Krypton lasers produce beams having many different wavelengths, all discrete and monochromatic, but capable of being split into colors.

The laser produces the picture because the human eye sees a spot for about 0.05 second, even if it is flashed for just a tiny fraction of that length of time. If the spot moves in such a way that it retraces the same pattern over and over at a rate of one complete cycle every 0.05 second, or faster, then the eye will see the image as a complete picture rather than just as a fast-moving spot. Of course, great intensity of light is needed for this to make a visible display, and the more complicated the pattern, or the larger the picture, the more intense the laser beam must be. There is a practical limit to the complexity of laser pictures that can be formed, because of this constraint, and also because the mirrors can be moved only up to a certain speed.

Because of these limitations, it is not possible to make complex or fast-moving pictures in three dimensions using the laser scan-

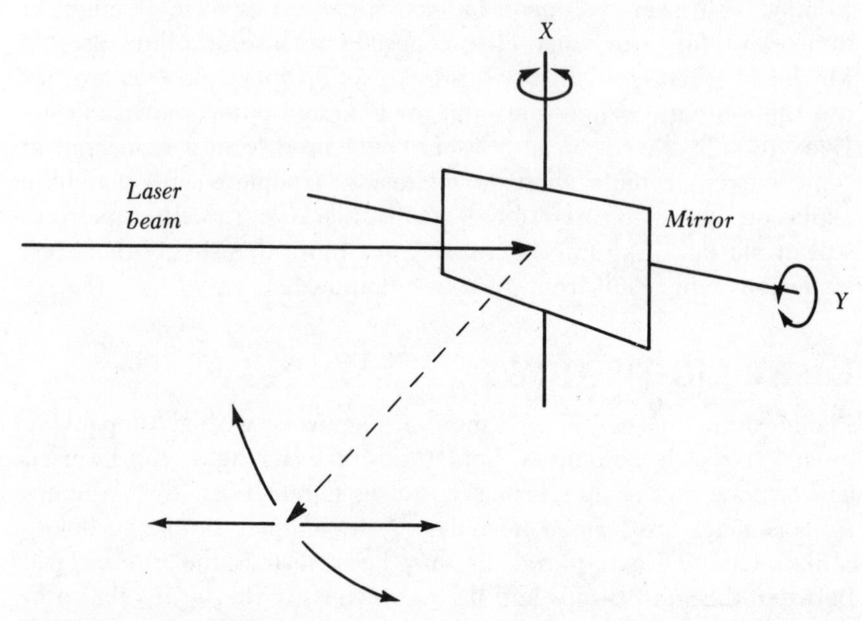

Fig. 5-13. A mirror, rotatable in two dimensions X and Y, can produce a fast-moving beam for laser shows.

ning technique. The electron beam that creates a television picture moves much faster, in terms of angular velocity, than does a laser, and this is mainly because the electron beam is deflected by electric and magnetic fields that can be varied much more rapidly than the position of a mirror by mechanical means. Laser displays of the scanned variety are therefore limited to line drawings. Complex patterns can be formed by moving the mirrors in ways that have certain relationships with each other, and the result is a geometric figure similar to (but more sophisticated than) the Lissajous figures you may have seen on an oscilloscope. Laser display devices can, in fact, be employed as oscilloscopes for relatively low-frequency disturbances.

Laser shows fall into the category of "kinetic art" since the images produced are usually moving images, changing position, size and shape. A musical accompaniment provides the entertainment of audio and video at the same time and has a mind-clearing effect because of the overwhelming of the hearing and vision at the same time. A planetarium provides an excellent environment for a laser show since the dome covers about 50 percent of observable space for the audience. Motion can be portrayed on a grand scale in this setting.

Another method of producing images by means of the laser is the giant hologram. A hologram is a three-dimensional rendition of an image, rendered by the interference of light beams from different lasers. With the proper diffusion in the air, large moving three-dimensional images are produced, such as space ships hovering over an outdoor stadium at night. Such images can be combined with live music and the result is a spectacular show that, if not spoiled by bad weather, draws large audiences and vast sums of money. We will examine how three-dimensional images are reproduced in the next chapter.

6

Three-Dimensional Photography

Ordinary photographs have a singular limitation: they cannot realistically reproduce a three-dimensional scene. A photograph may give excellent rendition of spatial depth by means of the perspective principle—objects far away look smaller than, and are eclipsed by, nearer objects—but true recording of three-dimensional information is not possible using an ordinary camera. Not, that is, unless special techniques, using special lights, are employed. Lasers provide a means of obtaining true "3-D" photographs because of the coherent nature of the light emitted.

Holograms can be produced on two-dimensional film or they can be rendered in three-dimensional space as a true spatial image. In either case, three-dimensional information is stored. Some holograms produce different images as they are viewed from different angles, such as the little flirt who blows you an ethereal kiss from inside a glass cube as you walk by—and undoes it again if you come back from the other direction. Holograms also have pragmatic as well as artistic uses. For example, some credit cards now have tiny holograms on their faces so that they can be uniquely identified. This makes it more difficult to commit fraud using the cards. Holograms can be used in entertainment; they complement laser shows and can even be used to play jokes, such as re-creating the inventor of the hologram, Dr. Dennis Gabor, in an office at CBS Laboratories. Holograms are used to inspect parts for machines and to scan grocery product codes at the supermarket.

THE DEVELOPMENT OF HOLOGRAPHY

Three-dimensional photographs existed in a sense before holograms. A "stereo" still picture could be produced by combining two photographs, taken from slightly different directions, and having the left eye see the "left-hand" photograph and the right eye see the "right-hand" photograph. The retinas of our eyes are two-dimensional, and it is primarily by parallax that we perceive depth (FIG. 6-1). This illusion can be re-created or even exaggerated using two ordinary photographs and a special combining scope. However, these "stereo" photographs cannot give the impression of eclipsing, where objects hidden behind other objects will come into view when the observer moves. The vantage point is fixed. The pictures can be put into motion by running two simultaneous movies, but willful changing of point of view by the observer, with full rendition of all parts of the scene, is not possible unless special techniques are used.

Holography was originally conceived as an improvement in the electron microscope, a device that is used for viewing very tiny objects at high magnification. A simplified pictorial diagram of an electron microscope is shown in FIG. 6-2. The focusing lenses in the electron microscopes of the 1940s were so poor that Dr. Gabor decided to try to bypass the lenses entirely. The problem with the lenses was in the focusing. Images that were blurred were not resolvable. If there were some way to accurately record the direction and intensity of light from an object, then all the information needed to re-construct the image would be on film. A visual image consists of nothing more than a field of view, in which the intensity of light from any given direction is represented by a point of variable brightness. More accurately, this is the technical rendition of a black-

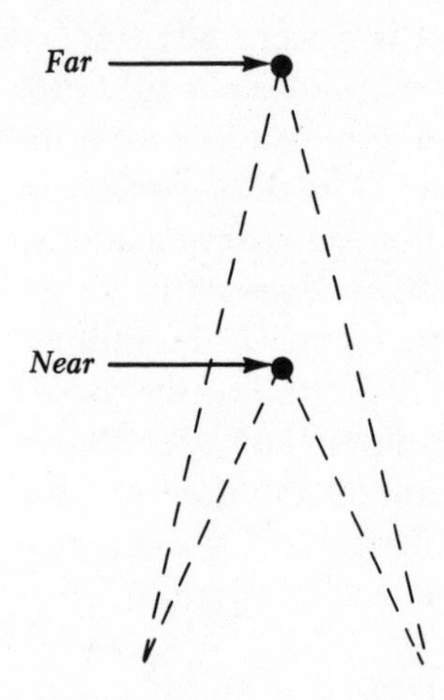

Fig. 6-1. Parallax results in different images as seen with the left eye (L) as compared with the right eye (R). Here, the far object appears on the left as seen with the left eye and on the right as seen with the right eye.

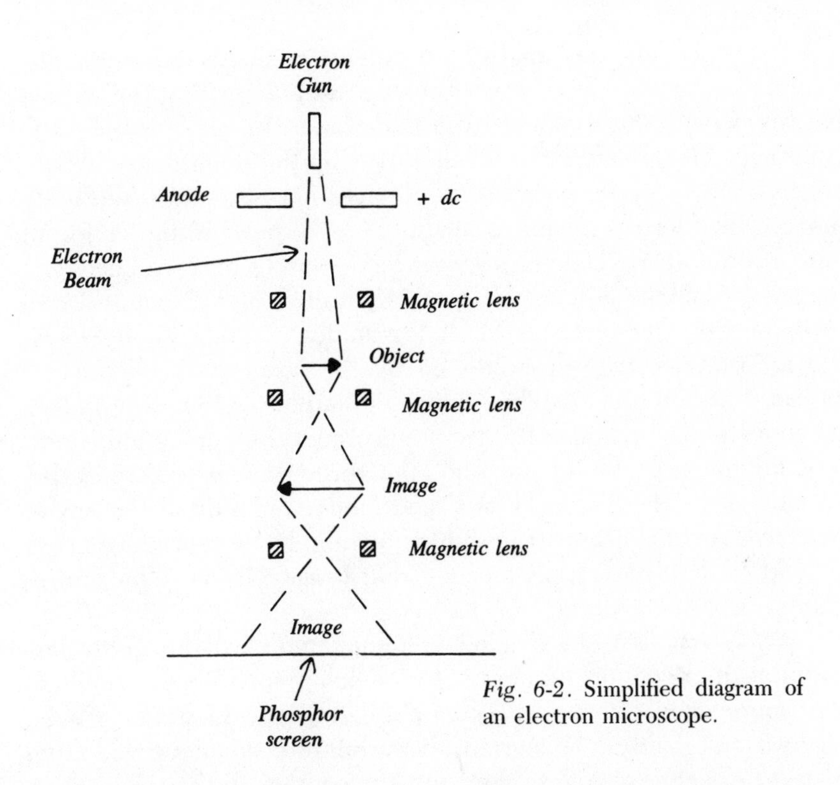

Fig. 6-2. Simplified diagram of an electron microscope.

and-white image. The photograph containing all of the necessary information, Gabor thought, might look like junk if viewed in the ordinary way, but if the information could be retrieved using special apparatus, the result might be superior to the electron microscopes then in existence. Optical means could be used to "correct" the electron-produced image, gleaning the intensity and direction information from the film.

When the light from a scene, such as a city skyline for example, hits a photographic plate without passing through any lens, the resulting "picture" is plain gray, without apparent detail. No picture results when a wavefront strikes a photographic plate; the exposure is essentially the same everywhere, resulting in a uniform shade of gray if the film is black and white. But if a reference beam, emanating from a point source and having coherent wavefronts, interferes with this light, the image can be developed into a picture. When this picture is observed, it is just as if the scene were being observed as it originally was, including depth, so that objects hidden from view when seen from one point may emerge when seen from another. The whole scene is recorded and—in Greek *holos gramma*, or the "whole message"—can be recovered, even when the film itself is only two-dimensional.

This discovery was made by accident, as many important discoveries are made. Gabor was probably not anticipating the impact that his investigations would eventually have. In fact, for a decade and a half, his discovery was passed over by the scientific establishment. This was partly because the devices necessary to make high-quality holograms were not available. Lasers provide the coherent light and the nearly geometric-point light sources that are necessary to make excellent holograms. Juris Upatnieks and Emmett Leith were the first to use lasers for holographics in the United States. The interference pattern produced by the laser-generated images appeared on the film as diffuse blobs or smudges when observed in the conventional manner. But by duplicating the illumination conditions for the exposure of the film, the three-dimensional rendition was obtained. The discovery was made independently in the Soviet Union by Yuri Denisiuk. He used his techniques to reproduce art so it could be enjoyed by people not able to actually visit the museums.

When you first see a good hologram, you will be quite surprised at the detail of the image. Color holograms seem at times to have more vivid colors than color photographs, and seem to even improve on "reality." Holograms look different when viewed from different directions because they actually contain all the information necessary to produce accurate images from various directions. In this sense a hologram is a kind of "super-real" image. It has all of the information that actually exists in the objects of a real scene, even more than you would obtain by looking at the scene itself, since the hologram is time-independent. It stays the same wherever you view it from, and this is not generally true of real scenes unless there is no motion.

IMAGE INFORMATION

To understand how holography works, the first concept is that of image information. A scene can be faithfully reproduced by recording the images as seen from a variety of different points of view. This need not include every possible vantage point in three-dimensional space, but the greater the number of points of view, the more complete the image information. A true three-dimensional scene is the combination of two-dimensional images obtained by looking at that scene from every point in space (FIG. 6-3). The true hologram, of course, contains the information from only a small portion, but still, in theory an infinite number of these points in space. This can be done on a two-dimensional continuous medium because, mathemat-

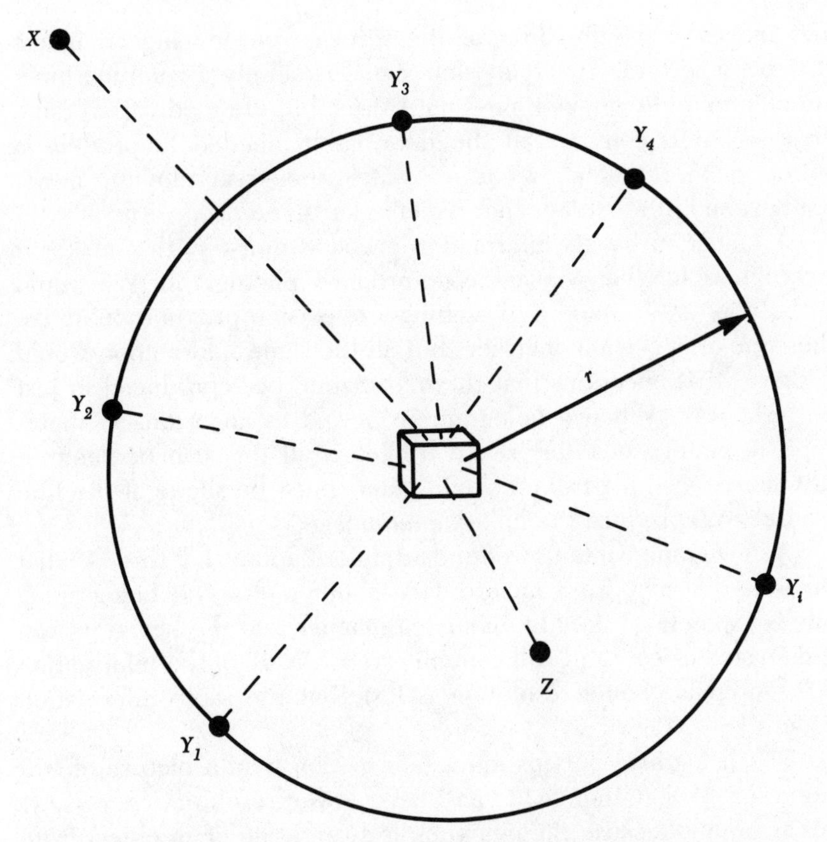

Fig. 6-3. An object has a unique image for every point in space from which it is viewed. Here, point X is outside a sphere of radius r. Points Y_1, $Y_2, \ldots$, Y_i, are on the sphere. Point Z is within the sphere. All possible directions are covered by the points Y_i but even all of these do not account for depth.

ically, there exists a one-to-one correspondence between all the points on a film surface and all the points in a given section of three-space. We state that here without proof, as that would be an exercise more suited to a text on math. It is intuitively apparent, however, if you think of the notion that infinity is equal to itself—an oversimplification for the pure mathematician, but adequate here!

If a picture is worth a thousand words, it is quite appropriate to say that a hologram is worth a thousand pictures. In fact a hologram is a combination of many thousands of pictures, put together so that the mind can make three-dimensional sense out of it.

A photograph is a compilation of information, the simplest being coordinate position and brightness. This is how a black-and-white television works. A spot, produced by an electron beam, scans a plane surface in a precise pattern. The brightness of the spot varies as it moves along the lines of the screen, reproducing an image 30 times a second. The television signal is simply a complex function of amplitude versus time—an AM (amplitude-modulated) radio carrier—but it contains all the information needed to produce a motion picture that allows us to watch the great movies, news, weather, and other things that we take for granted.

A hologram has its information encoded in a way that makes it unrecognizable when viewed as an ordinary photograph. We would not be very entertained by watching a television program on an oscilloscope or spectrum analyzer, but all the same information would be there. It is necessary that the information be reproduced in just the right way. When a hologram is viewed as an ordinary photograph, a pattern of interference fringes is all that can be seen. A light source, coming from a single point, must be shone at the film in order to reproduce the holographic image.

A hologram contains redundant information and this is what makes it different from an ordinary photograph. The holographic code is exposed on the film in such a manner that the hologram can be damaged or cut, and still contain essentially all of the information of the original. Some resolution is lost, but the same information remains.

The hologram contains more information than a picture of two dimensions, even though it may have lower resolution per view. This is simply because the hologram is a rendition of an object from many different viewpoints. True perspective is therefore possible. In FIG. 6-4 this is illustrated in three views of a cube. With a conventional photograph, only one of these views would be obtained. With the hologram they are all contained in the film, as are all of the possible intermediate views from within a given section of space. This section of space, the set of all points of view from which the cube may be seen, is generally a part of a large sphere surrounding the cube, or whatever object is exposed during the making of the hologram. However, the hologram may contain other points of view—for example, they may not all be from the same distance, so that an object will appear to move closer or get farther away as a hologram is observed from different points. Motion may be recorded so that the subject appears to move, for example, to blow you a kiss, as you view it from different angles.

Looking at a hologram is something like looking at a sophisticated computer graphics display. In fact, these two images represent similar degrees of improvement over conventional photographs and cathode-ray-tube displays.

Holograms were originally designed with the microscope in

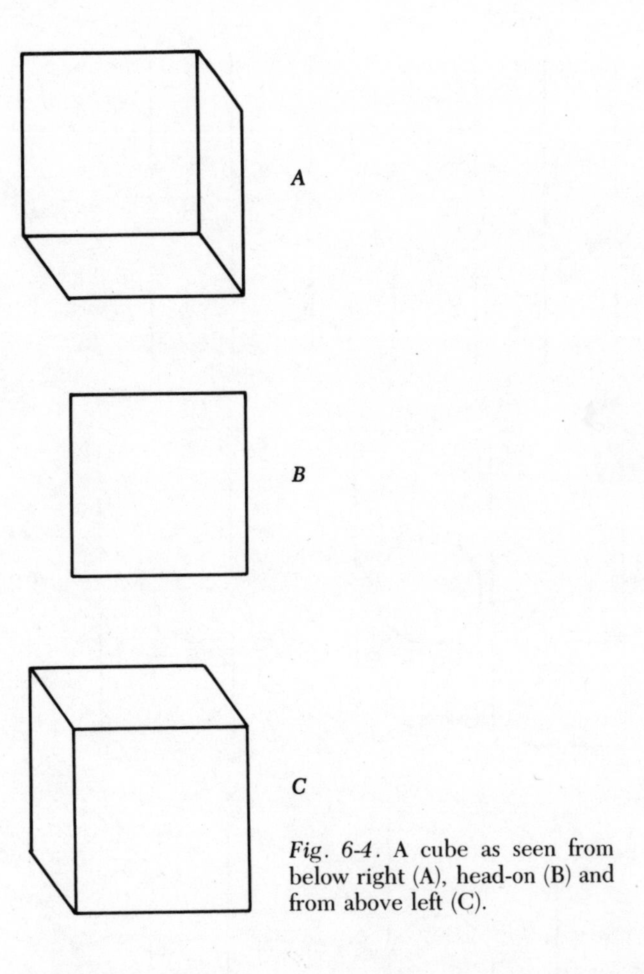

Fig. 6-4. A cube as seen from below right (A), head-on (B) and from above left (C).

mind, and holograms are used with microscopes to get more detailed images, rendering them in three dimensions rather than the common two-dimensional views. Holographic microscopy has been used to determine the velocities of tiny particles ejected by burning fuel. More detailed views of bacteria and other microorganisms have been obtained by photographing them using holographic methods.

WAVE DIFFRACTION AND INTERFERENCE

Holograms make use of wave properties of light called diffraction and interference. Diffraction is the property of a wave disturbance that allows it to bend around corners, especially those that are much smaller than the wave itself. Interference is the combination of waves in such a way that they alternately add and cancel in magnitude.

Perhaps you have seen diffraction in waves on the surface of a

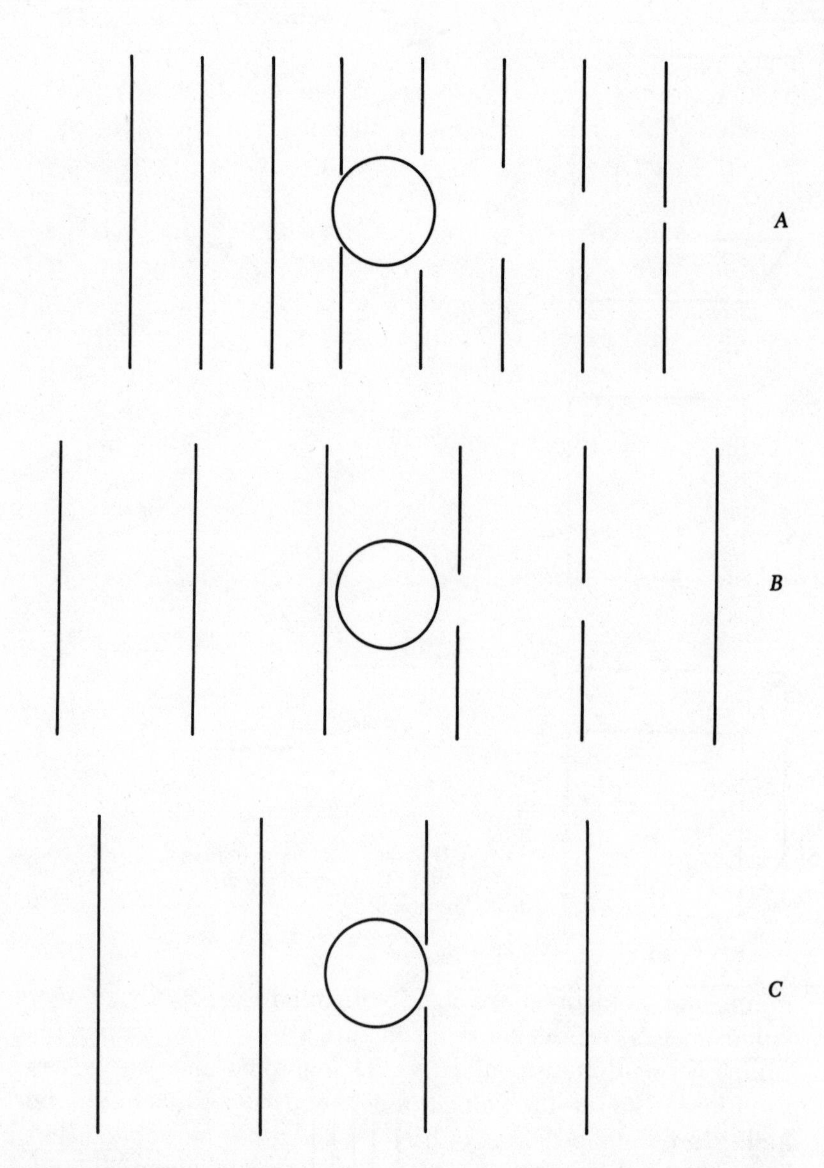

Fig. 6-5. Wave trains are affected less and less by an obstruction as the wavelength increases. (A) The wavelength is somewhat less than the diameter of the obstruction. (B) It is slightly greater and (C) much greater. The "shadow" becomes shorter and shorter, and is almost gone in C.

body of water. Scientists use a ripple tank, or a tub of water with wave generators, to examine the properties of wave disturbances, since water waves behave in many ways like sound or electromagnetic waves. You may have noticed that waves can bend around sharp corners, or pass by poles in the water (like the stanchions on the dock at the lake cottage). The corner or pole or other object, as long as it is small in size compared to the wavelength, has little effect on the wave train. The "shadow" produced is very small, getting smaller as the wavelength increases (FIG. 6-5). This effect occurs for sound waves in air, and also for electromagnetic waves. This is why, for example, you can hear someone speaking to you from around the corner of a building outdoors, and why cities do not have much effect on the propagation of low-frequency radio signals.

When the wavelength gets shorter, "shadows" develop. But the edges of these shadows are never perfect lines, even if the source of the disturbance is a perfect geometric point. There are interference fringes around any shadow when the source is nearly a point. These fringes disappear if the light source is large because of blurring effects. You may have noticed this effect in sound or seen it demonstrated in a ripple tank, but here we are concerned with this effect in the visible-light part of the electromagnetic spectrum. The interference fringes are the most pronounced with a monochromatic, point source of light, and the best source for this purpose is the laser. The diffraction around the edge of the object produces a pattern—not enough to eliminate the shadow, but enough to create tiny fringes of alternating light and dark (FIG. 6-6). These fringes are so small that it is usually necessary to have a magnifying glass or microscope in order to see them.

Diffraction is also produced when light is shone through a tiny hole or slit. Through a hole, a concentric pattern of fringes is produced (FIG. 6-7). Through a slit, the pattern is parallel lines, alter-

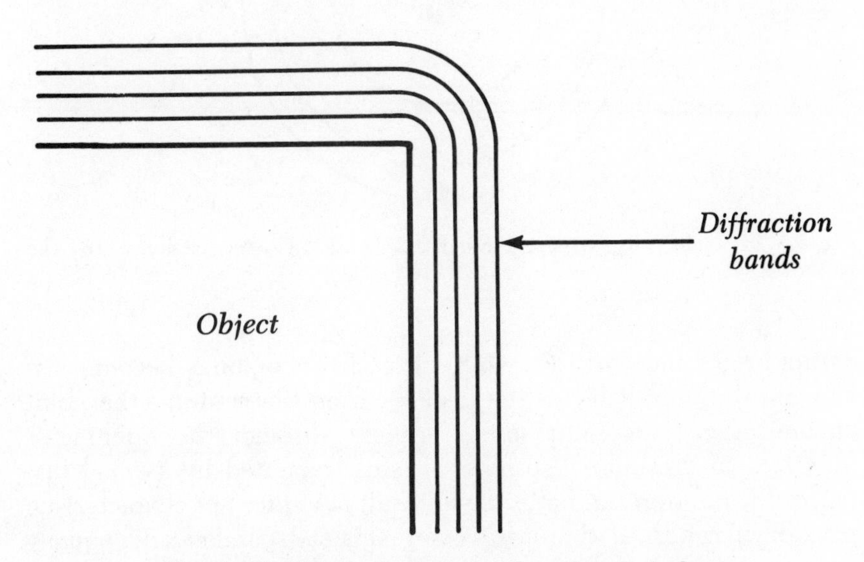

Fig. 6-6. Diffraction fringes around an object's shadow.

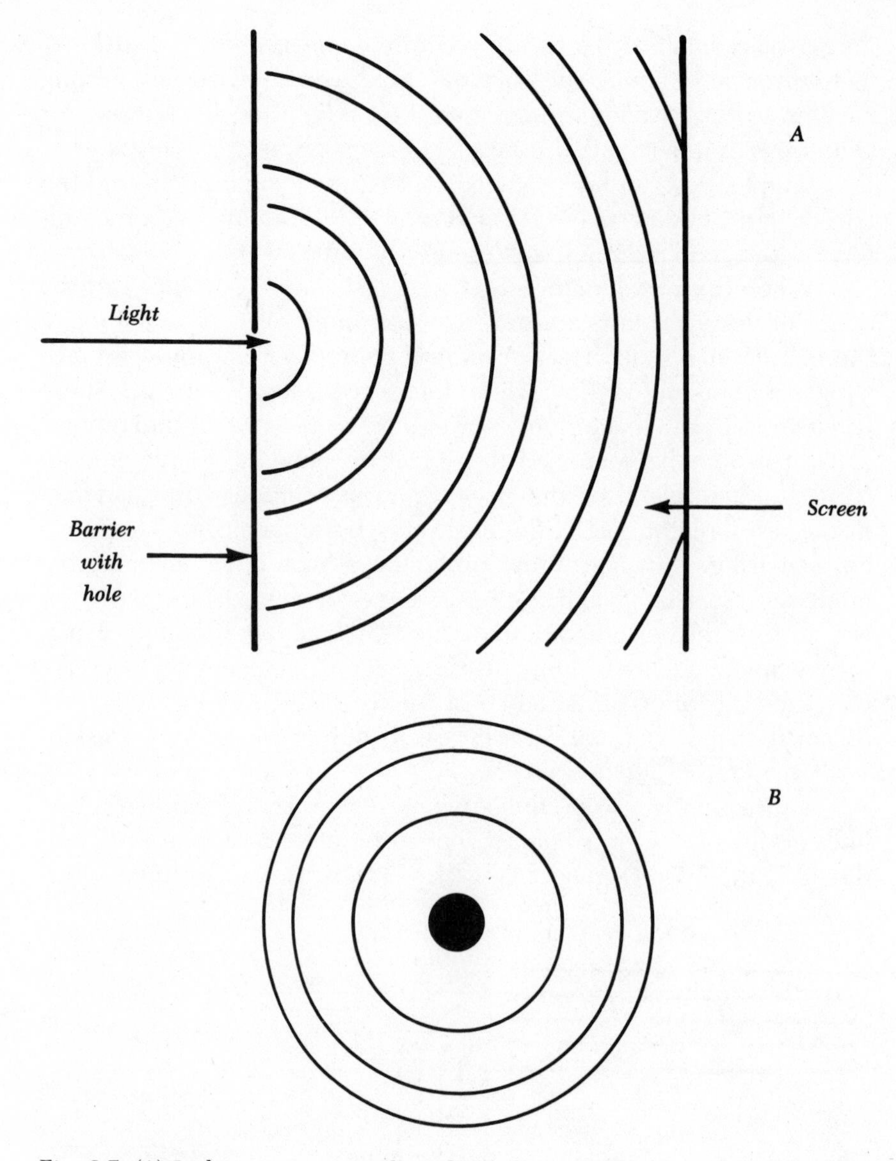

Fig. 6-7. (A) Light waves passing through a hole are diffracted as shown. (B) The pattern is shown.

nating bright and dark (FIG. 6-8). The circles or lines become less and less distinguishable as the distance from the center—the point or line where the light shines straight through the opening—increases. If there are two holes or slits separated by a small distance, the interference pattern will result in a different characteristic pattern of bands. If there are many slits, all parallel and equally spaced, the result is similar to that of a prism. Different wavelengths

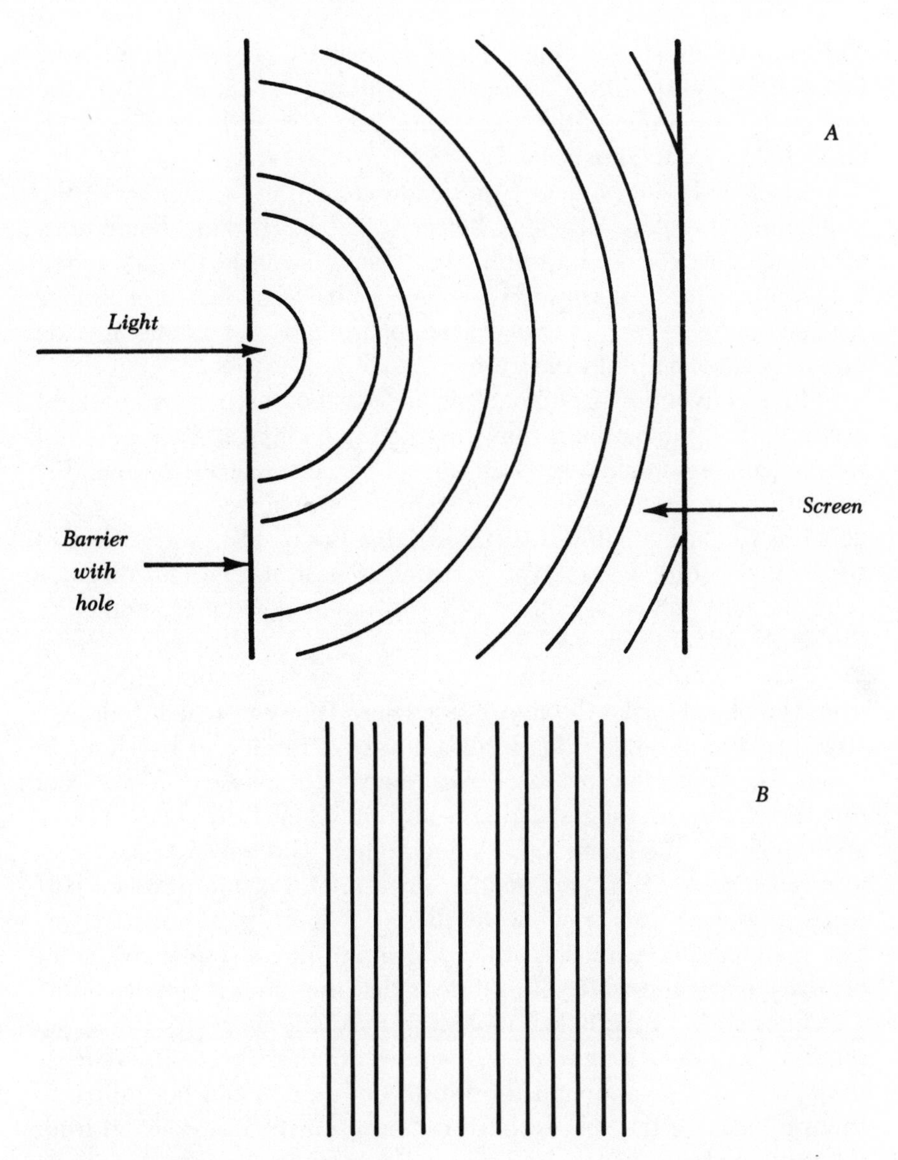

Fig. 6-8. (A) Top view of diffraction occurring from a slit; (B) the characteristic pattern of fringes.

of light will be diffracted differently, and the colors will be spread out into a spectrum if the original light is white. Thin films, with thousands of black lines etched on them, are used for this purpose and are called diffraction gratings. They behave just like prisms, except that the principle is diffraction rather than refraction.

Interference patterns in visible light can easily be recorded on photographic film. This is part of the process of making a hologram.

The sharper the interference patterns that can be produced by a given light source, the better the resulting hologram. Also, the higher the resolution of the film—the finer its grain—the better, as far as holography is concerned.

Black-and-white film is superior to color film in fineness of the grain, and therefore provides better detail for making holograms. Strangely, black-and-white film can actually be used to make color holograms. This is because the color results from the interference patterns of the waves. However, monochromatic holograms provide better detail than multicolored ones.

In a conventional photograph, a lens focuses incoherent light waves on the photographic film, producing the image. Each point on the image corresponds to a half-line in space, emanating from that point through the lens and out into space. Of course, the point that gets recorded on the film is that point that lies on the shortest line of sight outside the camera. The focusing is best at a certain distance from the lens, actually, within a certain range of distances known as the depth of field.

When a hologram is made, no lens is needed. The light waves from the object strike the film everywhere they can reach it along a straight path, and this is an infinite number of different half-lines in space. The depth information is recorded by the phase differences of the wavefronts as they strike the film at various distances from a given point on the scene to be holographed. The hologram is therefore not a record of a lens image, but a record of interference patterns produced from a coherent source of light, generally a laser. The photographic emulsion of the hologram film is similar to that of ordinary black-and-white film, except that the grain is usually finer. The finer grain of the holographic film makes it necessary to expose the film for quite a long time. Holograms are therefore still scenes as they are exposed, although the illusion of motion can be built into them by making the subject have different positions as viewed from different angles.

The finer the grain of the film, the better it is able to record the gray areas between interference addition and cancellation. The more shades of gray that can be reproduced in the interference patterns, the greater the detail of the hologram will be.

Fine grain also allows for resolution of more interference lines per inch than coarse grain. This allows for a wider viewing range, because interference fringes tend to get closer and closer together as the angle from the source increases. This effect is illustrated in FIG. 6-9 for a spherical set of wavefronts intersecting a plane film. The resulting pattern is a "bull's eye" with one light or dark center spot

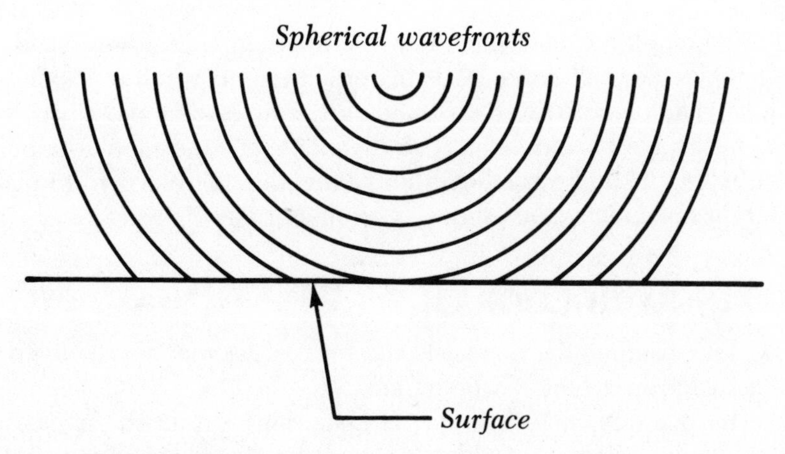

Fig. 6-9. Spherical wavefronts of light approach a surface and cause an interference pattern similar to a bull's eye.

and fringes that get closer and closer together with increasing radius (FIG. 6-10). Greater angles can be rendered with film having higher resolution, since more fringes can be reproduced.

The best films for recording the interference patterns necessary for making holograms are the silver-compound films. Some of these can resolve up to 10,000 lines per millimeter, or 10^7 lines per meter. This is hardly more than the actual wavelength of the longest red visible light. Dichromated gelatin holograms are also very efficient. This substance is sensitive primarily to light in the blue and green parts of the spectrum. This makes it impossible to use a helium-neon laser since it produces red light. Dichromated gelatin holograms have vivid color and high resolution despite their disadvantages. The image is formed by shrinking of the material when it is dried.

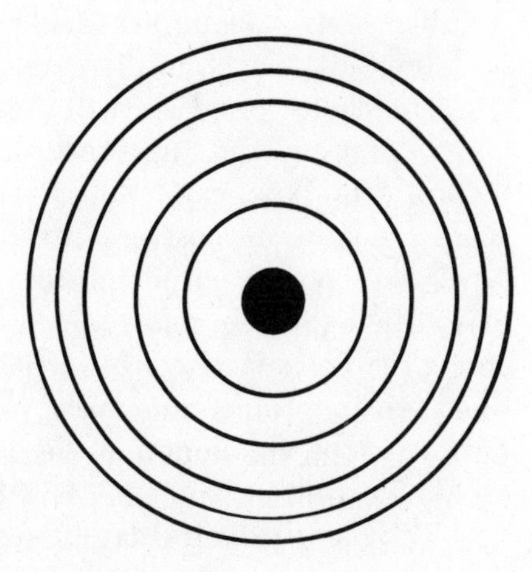

Fig. 6-10. The bull's eye pattern caused by wavefronts of FIG. 6-9.

Thermoplastic and photoresist are other techniques used to make holograms. Thermoplastic material deforms when it is exposed to heat. Photoresist material is eaten away by certain chemicals that form in the presence of light. The advantage of a thermoplastic hologram is that it can be used over and over, that is, reformed to make different holograms, sometimes as often as 1,000 times.

MAKING A HOLOGRAM

Now, let's examine the materials and techniques that are involved in the actual preparation of a hologram.

The main characteristic for all holograms is that the apparatus must be physically very stable. It cannot be allowed to vibrate. Motion or vibration among the different parts will upset the interference pattern, ruining the exposure. A wavelength of visible light is very small—just 7,500 angstroms for red light and 3,900 angstroms for violet light—and this makes stability important in the apparatus for making holograms.

Sheer mass is one method of obtaining the necessary mechanical stability. Objects with large mass have inertia and are less likely to be set in motion by minor disturbances. Cushioning and absorbent construction are two other ways to ensure that the holographic apparatus does not move or vibrate. Air cushioning is probably the best. A table may be constructed from heavy metal, such as steel and set on pneumatic supports. The heavy metal provides inertia and the pneumatic supports provide the cushioning. A basement is a better place to set up holographic equipment than an upper floor, since upper floors tend to vibrate more. Traffic on nearby streets can be a problem, but the work can be done in the middle of the night or the equipment can be set up in a rural area. Common plastic "bubble wrap" can be an excellent cushion, especially the variety with bubbles an inch in diameter (FIG. 6-11).

The choice of a laser will depend on available funds, the wavelength desired, and the power needed. Some lasers, notably the helium-neon type, can be obtained for as little as about $400. Other types are more expensive and there is no limit to cost. The helium-types have power output ratings from less than 1 mW to about 40 mW. The light from a low-power laser, or any laser, *is concentrated enough to make it dangerous if viewed directly*, so precautions must be taken to protect the eyes. Colored glasses that are non-transmitting to the wavelength of the laser will allow you to see what you are doing while not being at risk from stray laser beams.

The more powerful lasers are better than low-power ones only

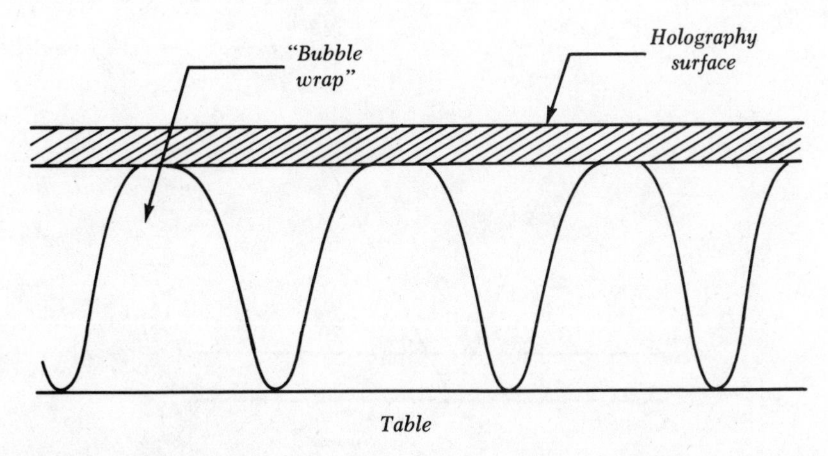

Fig. 6-11. Plastic "bubble wrap" can be used to cushion against vibration when setting up holographic apparatus.

insofar as the exposure time is decreased when the power increases. The higher-power lasers cost more and are more of a hazard to the eyes, but these disadvantages are offset by the greater flexibility they allow the holographer.

Multicolored images require the use of lasers of various colors. The most common are red, blue and green, the primary colors, so named because when combined with equal intensity, the result is a white light. Red, blue and green light can be combined to yield any color. This makes full, true-color holograms possible. Of course the cost is increased when three different lasers are used, as compared with only one laser.

Either continuous-output or pulsed lasers may be used to make holograms. The pulsed lasers are better suited to setups where there may be some vibration, since these lasers act like strobe lights, freezing the scene. The light is generally intense enough so that the entire exposure can be completed with one pulse. Continuous lasers generally cost less and have lower power, requiring longer exposures and a vibration-free apparatus.

First-surface mirrors are used in holography. The common reflecting mirror is silvered on the back surface of the glass (FIG. 6-12A) and will not work well because the front of the glass also reflects some of the light, creating a double image. The holographic mirror is silvered on the front surface (FIG. 6-12B), preventing this.

For high-resolution holography, collimating mirrors are sometimes used. These mirrors are concave, so that they have a focal point like a lens. The collimating mirror makes the wavefronts from a point source of light parallel and flat, rather than spherical.

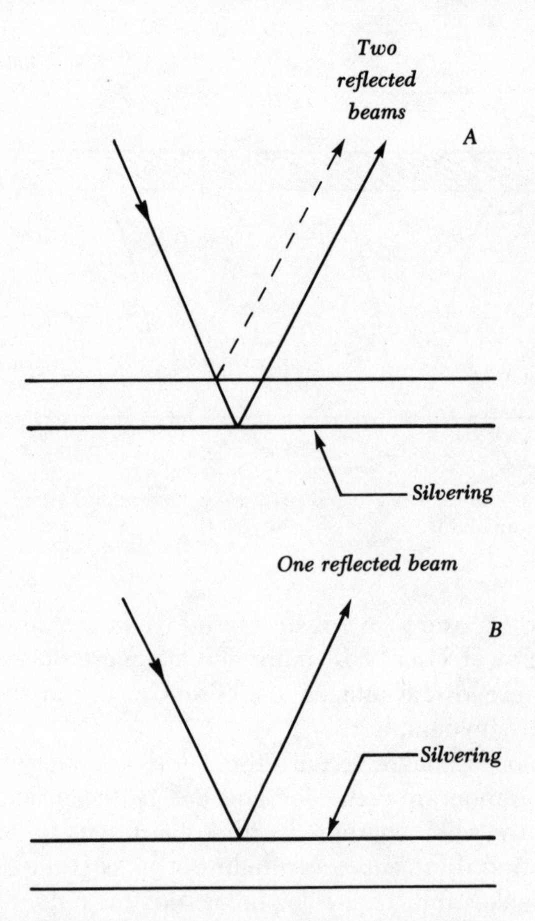

Fig. 6-12. (A) A typical mirror produces a double reflection. The first-surface mirror eliminates this problem, as shown in (B).

Lenses may be used to spread the laser beams out so that they illuminate larger areas than would be possible using a simple laser with its thin beam. Either convex or concave lenses will accomplish this beam-diverging function (FIG. 6-13). The lens spreads out the laser beam so it will illuminate the entire subject, while still having the necessary coherence for making sharp interference patterns on the film.

Holography equipment generally includes a beam splitter, which is a device that creates two laser beams from a single laser by a relatively simple process involving partial reflection. You have seen this principle at work when you look into a window and see not only what is inside, but also your own reflection. The simplest beam splitter is a partially reflecting, first-surface mirror set at an angle to the approaching laser beam (FIG. 6-14). Some of the beam is reflected and some passes through the glass unaffected except for a

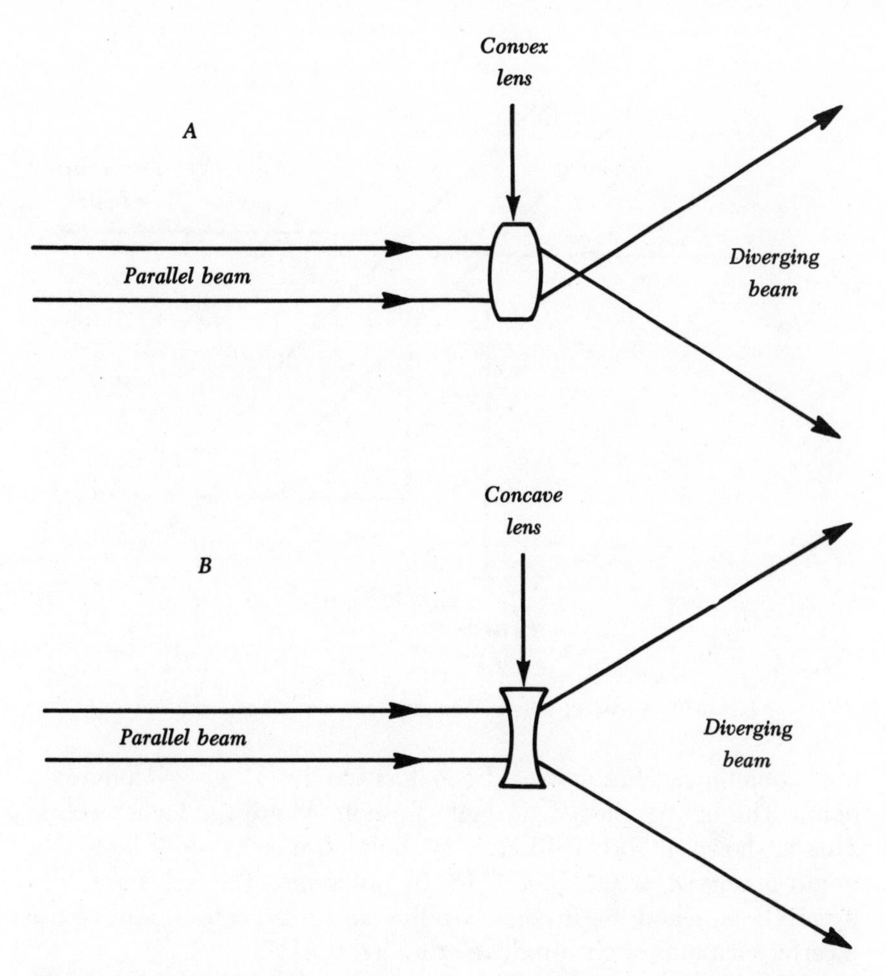

Fig. 6-13. Convex lens (A) and concave lens (B) used to spread out a collimated light beam for illuminating a subject.

slight diminution in intensity. A beam splitter may be silvered so that it splits the energy equally between the two beams, or it may be so made that the reflected beam ends up with more or less energy than the transmitted beam. In either case, of course, the sum of the energies of the transmitted and reflected beams is equal to the energy of the incident beam minus a tiny amount that is lost by absorption in the transmitting medium. Some beam splitters have more transmittivity over some areas than others. The ratio of transmission to reflection is often specified, with percentages totaling 100, such as 80:20 or 50:50. Some beam splitters can be adjusted so that the ratio varies and can be preset to a desired value.

The coherence of light can be improved using tiny "pinholes". The coherence is maintained by passing the light through the open-

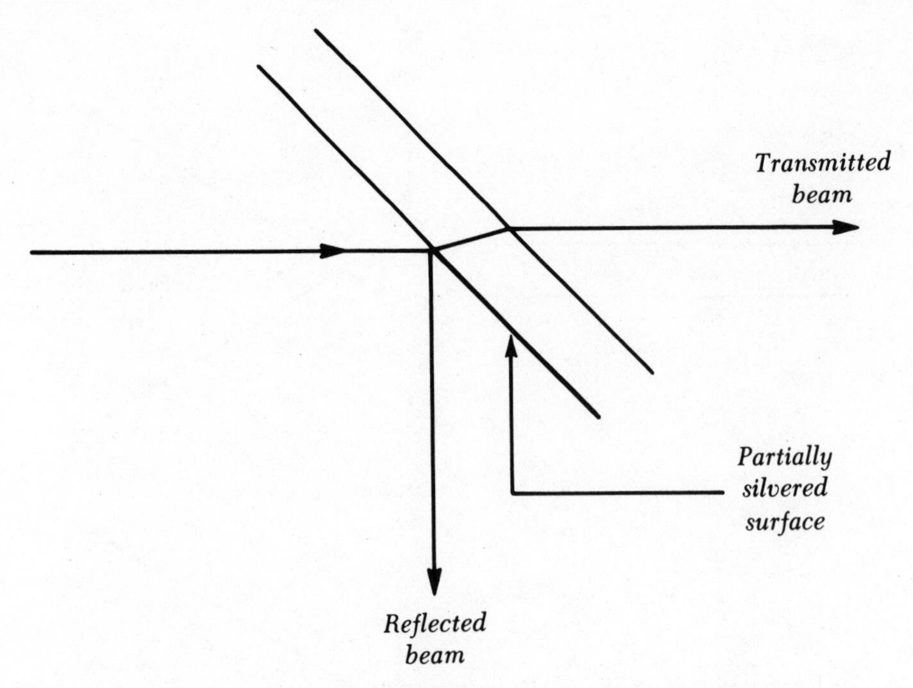

Fig. 6-14. A partially silvered mirror serves as a beam splitter.

ing, sometimes after having been focused by a convex lens to a point. The light is passed through the hole where the focus occurs. This is shown at FIG. 6-15. The pinhole eliminates stray light that might otherwise create "noise" in the hologram. The holes are very precisely punched; their edges are fine and regular to minimize the interference that might produce stray light.

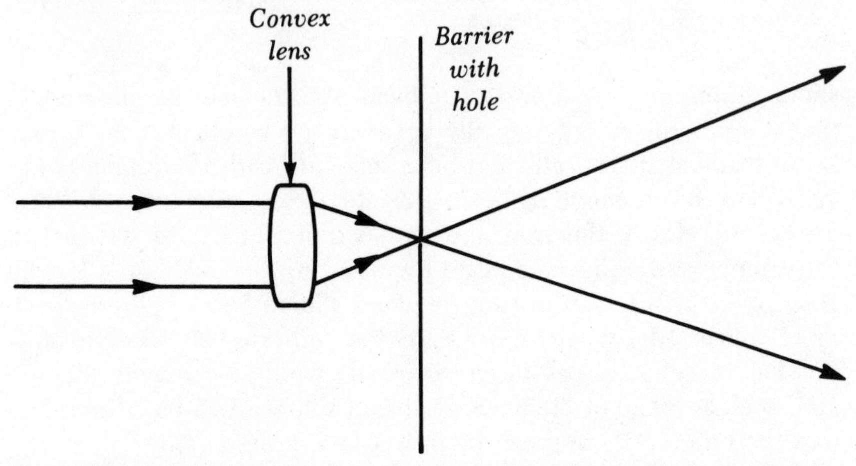

Fig. 6-15. A "pinhole" and lens are used to eliminate stray light in a holographic apparatus. The hole is at the focal point of the convex lens.

THE FIRST HOLOGRAMS

Gabor made the first holograms without using a laser, and this resulted in the necessity for a narrow angle between interfering beams. The narrower the angle at which light beams interfere, the coarser the fringes and the less detail is rendered. A mercury-vapor lamp, which produces bright light at discrete wavelengths, although the light is not coherent, was employed. The hologram was made by passing light through tiny letters. This caused the light to be diffracted, and this diffracted light interfered with the light that passed directly to the film without striking the letters (FIG. 6-16).

A more sophisticated early holographic setup is shown in FIG. 6-17. Here, a laser was used, and a concave diverging lens allowed the light to be spread so that it covered the entire subject. A mirror

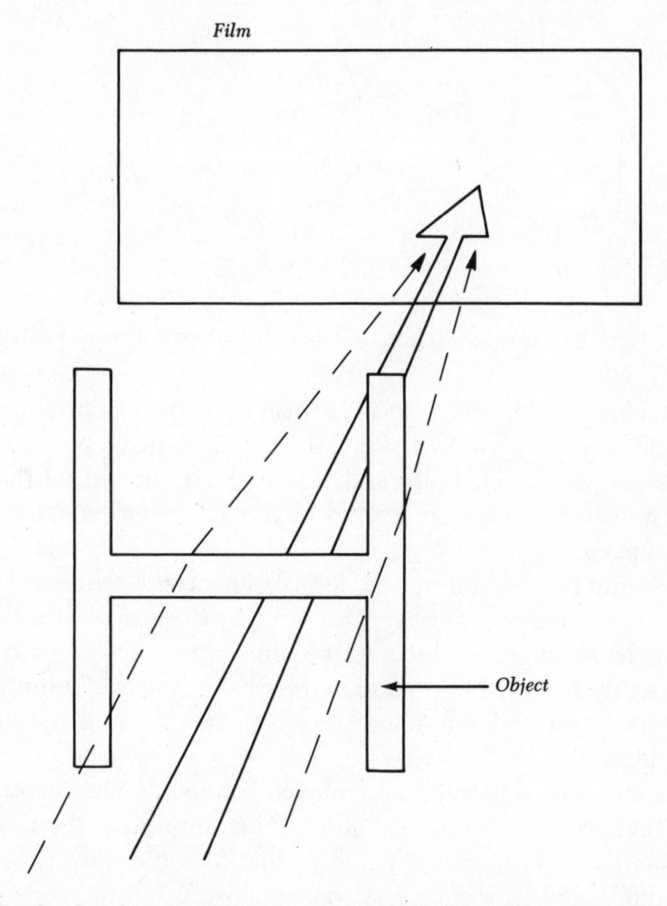

Fig. 6-16. The first holographic setup. The light beams, indicated by dotted lines and the solid arrow, were from diffraction and direct illumination respectively. The light source was not a laser.

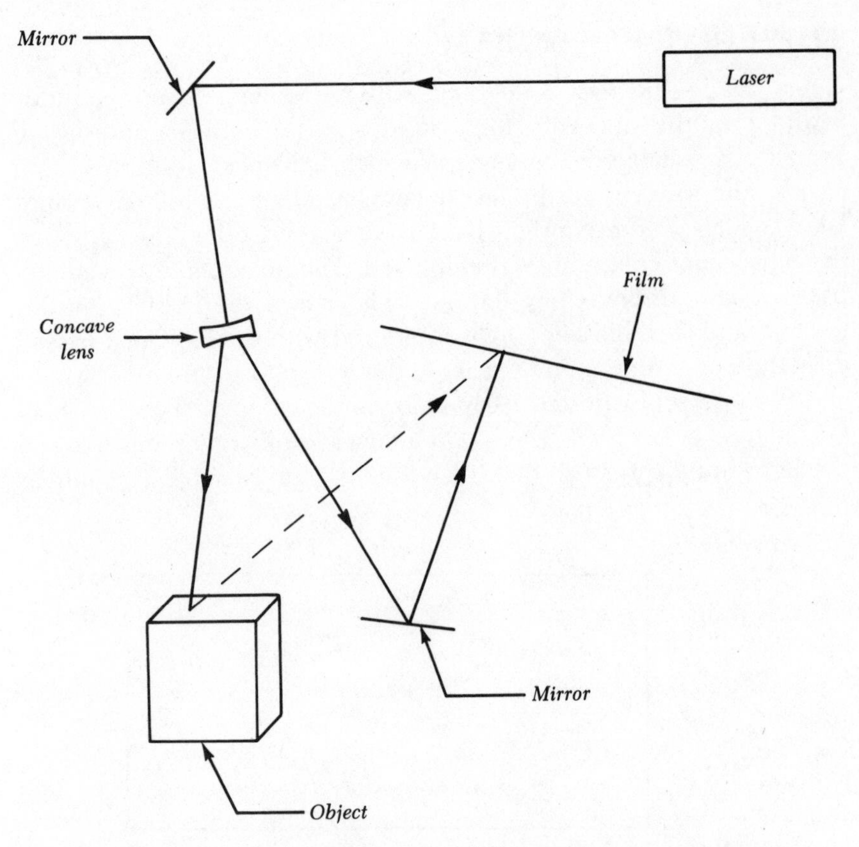

Fig. 6-17. A more sophisticated early holographic arrangement.

was used to provide the reference beam, or beam not affected by the subject. Note that this method still used just one beam—no splitter was employed. Leith and Upatnieks improved on this technique by using a beam splitter to obtain the reference and object beams separately.

By using beam splitters, the laser beams can be directed toward the film over greatly differing paths, so that they meet at a large angle. This large angle results in a fine interference-fringe pattern and consequently in greater detail. The relative strengths (intensities) of the object beam and reference beam can be adjusted by means of the splitters.

Ideally, the reference and object beams should traverse distances that are equal or nearly equal. This minimizes the difference in the amounts of wavelength change that take place over distances. If the difference in distance is too great, the interference pattern will not be clear and the quality of the hologram will suffer.

The greatest allowable difference in path lengths for the object beam and reference beam is called the coherence length. The path difference must not be greater than this amount or the quality of the hologram will be unacceptable. The mercury-vapor light originally used by Gabor had a coherence length of only about 1 millimeter. Modern holographic apparatus has a coherence length on the order of 1 meter or more. The increase in coherence length is largely because of the greater coherence of the light from the lasers, and also because of the greater intensity. The small coherence length of Gabor's equipment made it essentially impossible to keep the distance within the limit, so the quality of the end product was severely compromised compared with today's holograms.

REFLECTION AND TRANSMISSION HOLOGRAMS

Holograms may be categorized as either transmission type or reflection type. In a transmission hologram, the object beam and reference beam strike the emulsion (film) from the same side. In a reflection hologram, the object beam strikes the emulsion from one side and the reference beam from the other side. In FIG. 6-18, a typical transmission-hologram setup is shown. In FIG. 6-19, a reflection-hologram arrangement is shown. The reflection hologram is an improvement of the transmission type for several reasons, the most significant being that a reflection hologram can be viewed using ordinary white light. The reflection hologram often provides a greater range of views than the transmission type because the object is more evenly and completely lit. Y. Denisiuk of the USSR was the first holographer to use the reflection technique.

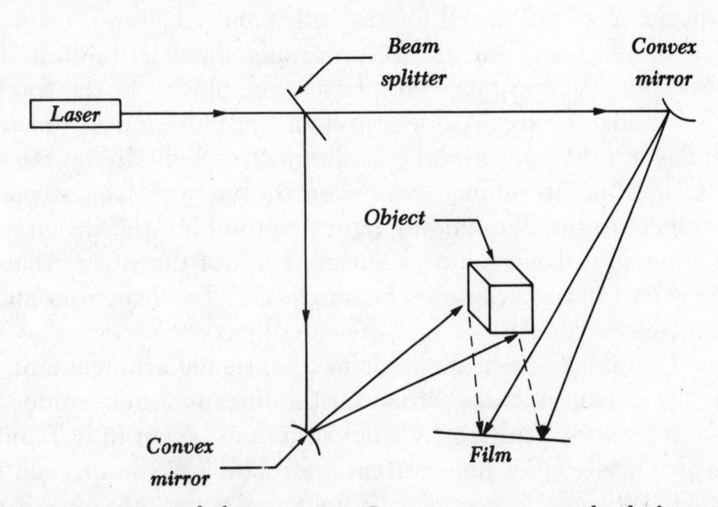

Fig. 6-18. A transmission-hologram setup. Convex mirrors serve the dual purpose of a flat mirror and diverging lens.

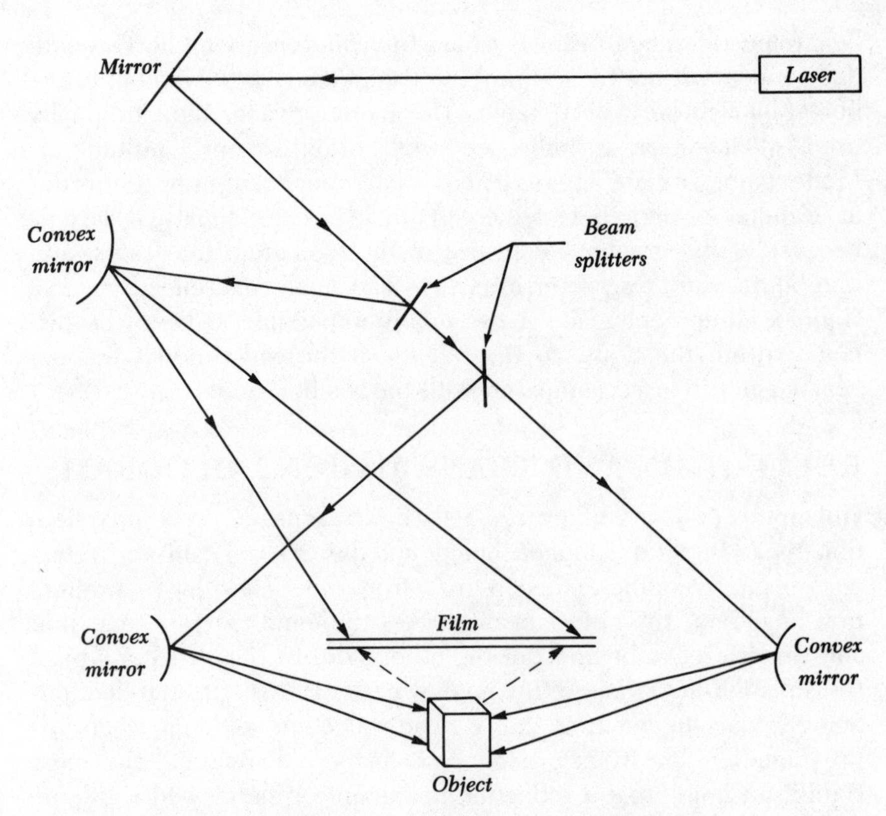

Fig. 6-19. Arrangement for making a reflection hologram.

The reflection hologram requires a film that can be exposed from either side. Ordinary photographic film is opaque on one side, so a special film is required for the reflection hologram.

To view a hologram, the setup arrangement is duplicated, except that the observer takes the place of the object. In the transmission hologram, the light passes through the film, and in the reflection hologram the viewer and the illuminating light are on the same side of the film. In either case, when the viewer changes position with respect to the illuminating light and the film, the image moves in position as if it were a real object. Parts of the image that cannot be seen from a given viewing angle may be clear from another viewing angle. Parallax occurs between objects at various distances, just as it would in a real scene or in the original arrangement. It is the angle, not the actual position on the film, that determines what the observer sees. This is why a hologram can be cut in half and still retain practically all of the position information of the original hologram. Detail is lost, however, as a hologram is cut into pieces, since there are fewer interference fringes in the piece than in the original.

There are other ways to decode, or demodulate, holograms, but they are too numerous to be thoroughly covered and illustrated here. A book solely devoted to holograms is recommended for the serious holographer-to-be.

Perhaps the best way to describe a hologram is to say that it is a modulated camera-film rendition of a scene. The method of modulation is just different from that normally seen in a photograph. But all the information needed to provide a realistic three-dimensional image is there, obviously, in the grains of the film. It is only necessary that the proper "demodulator" be used in order to reproduce the original image. We cannot listen to single-sideband radio signals without a special detector. Such signals are unintelligible using an ordinary envelope detector, but all the information is there and can be recovered with the product detector. Similarly, holography is not really magic; we are not cheating on two dimensions and creating anything by illusion or mathematical trickery. We simply use a different form of "modulation" to convey information that is already there and can, as such, be stored for future use.

7

Lasers in Warfare

Lasers can be used in a variety of military applications. There is a general preconception that lasers are deadly ray guns, capable of destroying tanks, aircraft, or even space ships at great distances and with deadly accuracy. This notion probably comes from several decades of conditioning by science fiction writers. It takes less effort to destroy things than to create them, and evidently less thought to envision mass destruction than sophisticated creativity.

It is true that some lasers have destructive power. But the uses of lasers in military situations goes far beyond just this raw energy. Lasers can be used to guide missiles, for communications, to blind surveillance cameras, and many other applications.

PHOTON TORPEDOES

You may have seen television renditions of such things as "photon torpedoes" and "phasers," both fictitious weapons that were used for disabling enemy space ships or even stunning or killing humanoids and other life forms. Both of these weapons were based on the concept of the laser. The photon torpedo was basically a high-energy burst of visible light. The phaser was similar to a laser, getting its power from the coherence of its emissions. Such weapons are today at least theoretically plausible; however, the power required for a phaser would necessitate a much larger device than could be held in the hand, and the beam would probably not be visible as it was shone at the subject. Moreover the device would be noiseless and the damage would be localized, such as a burn or scorch. Scenes of the victims haplessly falling unconscious or being vaporized entirely are unrealistic.

The appearance of some science fiction weapons resemble giant ruby lasers, with resonating crystals surrounded by helical energizers (FIG. 7-1). The power supplies are gigantic, sometimes consisting of a complete power plant for just one weapon. A laser of that power would need a substantial cooling system but could be developed if the funds were appropriated. Such a device might be capa-

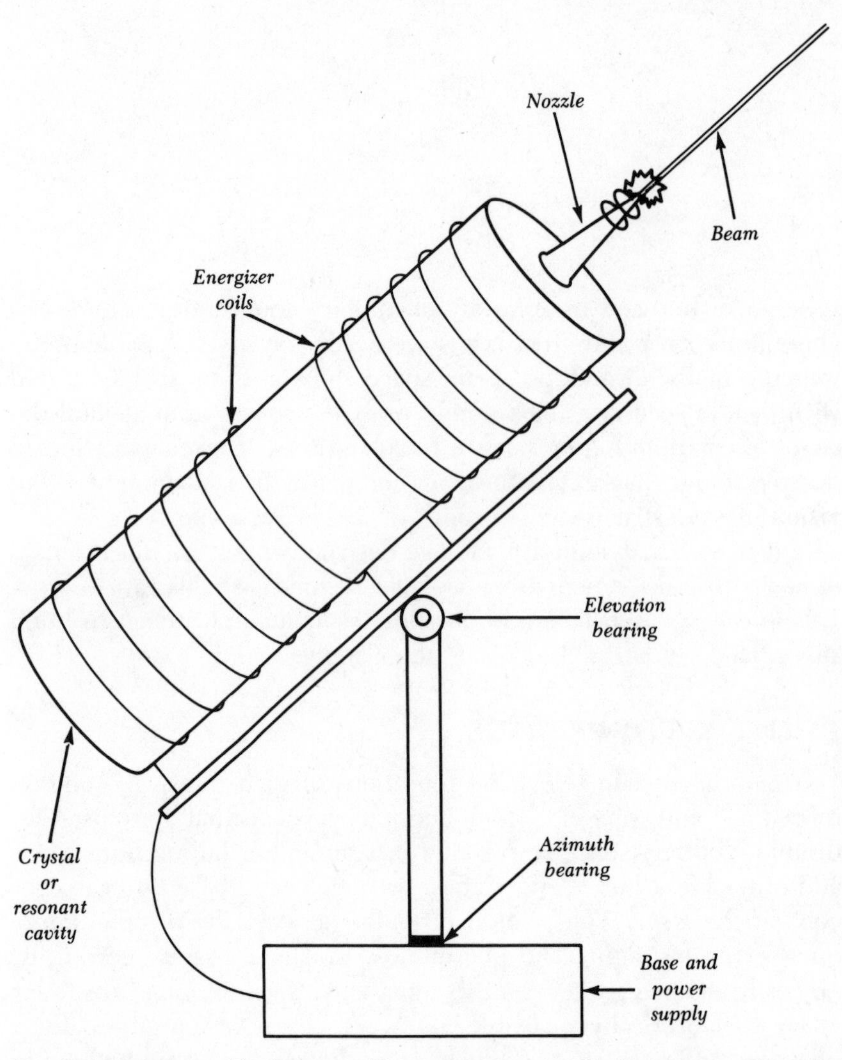

Fig. 7-1. Hypothetical high-powered laser weapon. Actually this is not an altogether unrealistic design.

ble of shooting down aircraft and satellites, but probably with no greater effectiveness than a nuclear or conventional missile. Gigantic, high-powered lasers would be better suited to long-distance space communications.

Photons cause heating of a target when the energy is absorbed and not reflected. Plain mirrors will cause most of the energy from a laser to be reflected, and a metallic, shiny object is therefore difficult to damage by means of a laser. Protection against the powerful photon torpedoes of science fiction would be simple: simply coat the spacecraft with reflective material.

Photons cause some force to be exerted on an object when they strike, and if this force were great enough, a whole spaceship might be literally "blown away" or off course. However, the power required to generate a light beam of this intensity is not within the reach of modern technology, and even if it were, the cost would be so great as to preclude its usefulness.

SHOOTING DOWN WARHEADS

Let us consider the difficulties of defending against an all-out nuclear attack.

Warheads would come primarily from intercontinental ballistic missiles (ICBMs). Some missiles would be launched from submarines, some from ships, and some from aircraft. Some aircraft might carry bombs directly to their targets. We in the United States have the advantage geographically that there are no foreign missiles that we can be certain of, yet, stationed in silos in this hemisphere, although some experts believe there are silos in Cuba. Each missile might carry several dozen warheads and at least as many "dummies," or empty warheads designed to make the overall picture more complicated from a tracking standpoint. In total, a full-scale attack might involve many hundreds or even thousands of individual warheads.

Anti-ballistic missiles (ABMs) might destroy enemy missiles before the multiple warheads were launched, but this is unlikely since the multiple warheads would, in many cases, be deployed before they came into range of our radar units. This has prompted scientists to look for early warning devices, such as over-the-horizon radar, to allow more time for launching accurate ABMs. Over-the-horizon radar is already believed to be in use by the Soviets. Its pulses can be heard in the high-frequency (shortwave) radio spectrum and the characteristic sounds have caused it to be dubbed "the Russian woodpecker." However, such a radar system does not provide very great accuracy. It may indicate that missiles have been launched, and even give a rough idea of their whereabouts. But the frequencies involved, and the interaction of the ionosphere with the radio signals, limits the accuracy of this kind of radar system and it would not be good enough for use by guided ABMs.

The sheer number of warheads, their speed, and the unpredictable nature of their movements, make for an almost impossible situation. Consider the scenario of FIG. 7-2, in which a single missile throws off several warheads and many dummies in all directions, and multiply this by, perhaps, 100 or 500. How would we ever destroy each one of these bombs? If there were 3,000 bombs and we got 99 percent of them, that would still leave 30 warheads, each with enough nuclear energy to destroy a city the size of Pittsburgh or Dallas. And we would almost certainly never achieve a success rate of 99 percent. It is a gloomy picture. Only a massive computerized defense system could begin to defend against offensive weapons already in existence and ready for use any minute.

The unpredictable movements of the warheads—even the attacking country would not be aware of where all of them would go—is made all the more difficult to track because of the enormous speeds involved—up to 10 or even 15 kilometers *per second.* At such speeds a timing error of just 0.01 second would result in a position error of 100 to 150 meters—more than the length of a football field. A laser with a beam just a few centimeters wide would have to be very well-placed in order to have the desired effect. And, the warhead would have to be exposed to the radiation for a sufficient length of time—several seconds at least—for destruction or disarming to take place.

If lasers are used as the means by which warheads are to be dis-

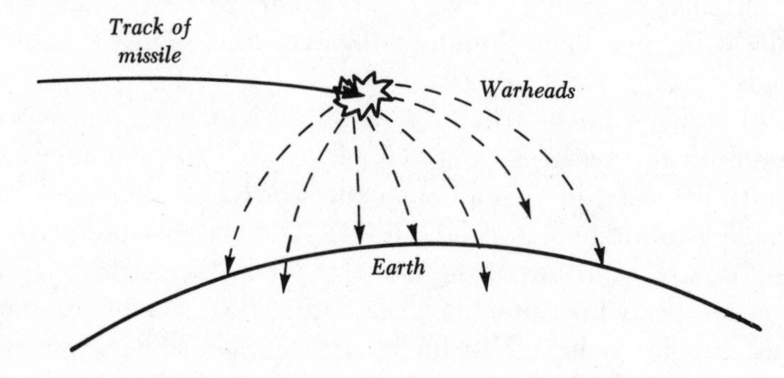

Fig. 7-2. A typical long-range nuclear missile has multiple warheads that travel in a variety of different paths, making it difficult to hit them all once they have been ejected from the main rocket.

abled, these devices may be located on the surface of the earth, either at ground stations or on ships; or they may be aboard aircraft or on manned or unmanned space ships. However, ground-based lasers would be the easiest to put together for several reasons. Primary among these considerations would be the simpler aiming apparatus, for only the motion of the warhead with respect to the Earth itself would have to be taken into account. Power supplies could also be made very large since they would not have to be hauled around. The limitations would be the fact that only those parts of the sky actually visible on a line of sight from a given ground station could be covered with the laser: the atmosphere would create attenuation, even in the absence of clouds, and this attenuation would become considerably more of a problem at lower firing angles (FIG. 7-3) because of the greater amount of air through which the beams would have to travel. Cloudy or foggy weather would preclude the use of optical lasers for shooting down warheads speeding high above the lower atmosphere.

The other excellent place to locate a laser weapons system would be on a satellite or, even better, on a space station. The United States has plans at the present time for building a substantial space station. A space station could utilize solar power, and could be placed in an orbit with exactly predictable and calculated paths, so that as far as a sophisticated computer might be concerned, the station could just as well be stationary on the surface of the earth. Numerous stations, in various orbits, could catch missiles early, just as

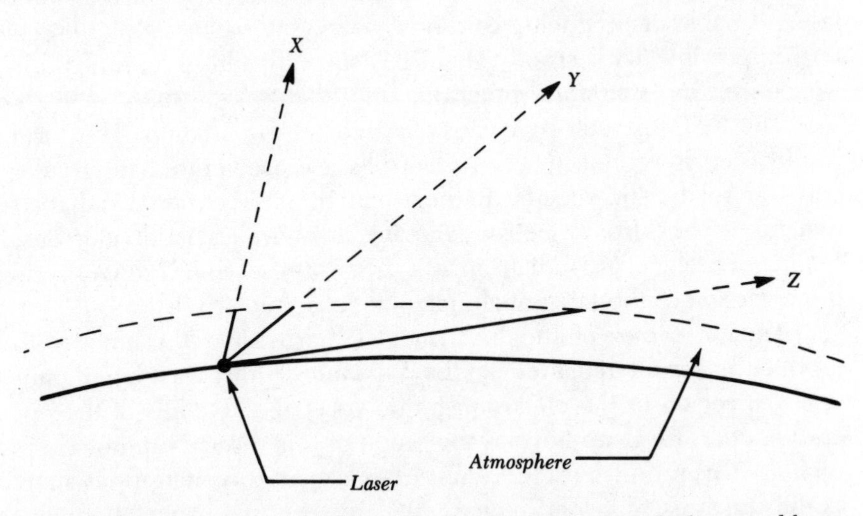

Fig. 7-3. Lasers must pass through more and more atmosphere for lower and lower firing angles. Here, beam X passes through the least air and suffers the least attenuation; Y and Z pass through more air and are absorbed more.

they were launched. It might even be possible to conduct a "pre-emptive defensive strike" and destroy enemy missiles while they were still in their silos. But this smacks of the intention of gaining the upper hand too firmly, of giving us the power, in theory at least, to blackmail the opposition. Laser weapons on spacecraft would have the additional advantage of not being limited by atmospheric attenuation, once enemy missiles had gained sufficient altitude.

Detection of enemy missiles would probably have to be carried out by visual means. Any other method, such as infrared detection or radar, is subject to jamming and the use of decoys of low cost. For a real visible image to be a decoy, it would mean the costly devotion of a whole rocket to nothing but empty nose cones—it would be less expensive to load them with real warheads. If a visual detection system, based in space, were extensive and sophisticated enough, the cost of overcoming it by sheer numbers (and that would be the only way) would be so great that any superpower would think twice. Space stations, moreover, could also be devoted to a wide variety of peaceful applications, in addition to the inclusion of defensive weapons.

It may sound strange to casually discuss ways to destroy weapons with other weapons being built to save humanity. There are some more practical results to all this weapons research, however.

FREE-ELECTRON LASERS

The Free Electron Laser (FEL) is a relatively new, intense form of laser, capable of producing coherent energy at wavelengths heretofore impossible with lasers. The FEL shows promise as a possible weapon in the "star wars" program.

The FEL operates by means of synchrotron radiation. This form of radiation is produced when electrons are accelerated in circular paths at relativistic speeds. Rather than the conventional radiation that takes place from electrons moving at nonrelativistic velocities, high-speed electrons radiate in a sharp, intense cone, generally in the direction of the tangential instantaneous velocity (FIG. 7-4).

The principles of the FEL involve harnessing this energy to produce coherent radiation in the extremely short-wavelength microwave region of the electromagnetic spectrum. Details of the operation of an FEL are beyond the scope of this discussion; however, such devices produce useful coherent energy at wavelengths as short as 400 microns (0.4 mm or 4×10^{-4} m). Future expectations are for shorter wavelengths, perhaps down to 100 microns or less, in the far infrared. This will increase the potential of the free-electron laser

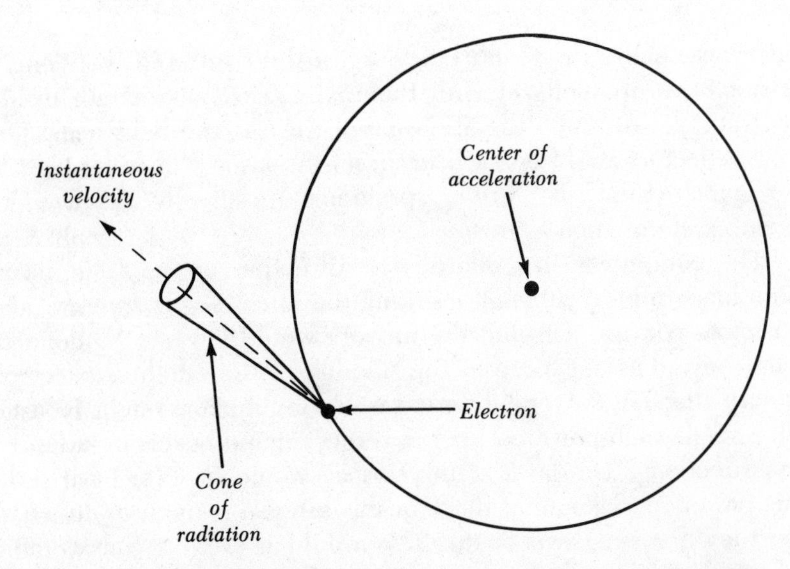

Fig. 7-4. The free-electron laser operates using the energy produced by high-speed orbiting electrons. This is called synchrotron radiation.

for generating "heat radiation." The FEL is a tunable device; that is, its wavelength can be adjusted at will for the desired application.

The FEL is a pulsed laser, emitting its energy in short bursts of extreme intensity. It is being looked at by solid-state physicists as well as medical professionals. Chemists and molecular spectroscopists are also interested in the possibilities of the shorter-wavelength FELs.

The FEL also can produce ultraviolet laser radiation. Applications of this radiation include high-resolution absorption spectroscopy in wavelength ranges not now possible. Industrialists are interested in its possible uses for depositing dielectric and metallic films, for example, in the manufacture of electronic capacitors and circuit boards.

Medical scientists are naturally interested in possible uses of any type of laser, the FEL in particular because of its ability to cause heating of objects it strikes. Microsurgery is the most obvious application of such a laser. Phototherapy, such as the heating of tumors and perhaps the irradiation of blood as is discussed in Chapter 3, is also a possibility. The carbon-dioxide and neodymium-YAG lasers are now commonly used for these medical purposes.

But the most publicized potential use of the FEL is in the Strategic Defense Initiative (SDI) program, where it is thought that the high power output and wavelengths in the infrared might make the device especially well suited for shooting down missiles. A high-powered FEL might produce a large amount of energy at 10,000 angstroms, in the near infrared. (Such a short infrared wavelength has not been attained yet with the FEL). This wavelength would penetrate well through the atmosphere, just as red light does. Orbiting reflectors could be used to direct the laser at missiles shortly after launch (FIG. 7-5). Military programs are already underway to develop systems such as this.

The computers that control the direction of the laser beam would have to be capable of orienting the mirrors very quickly, and by remote control. Turning the mirrors would have to be done in such a way so as not to upset the stability of the satellites to which they are attached. Several different reflecting mirrors might be used with a single high-powered laser to direct several beams in independent directions. The laser at the surface would have to have sufficient power to make all of these beams intense enough to do whatever they are supposed to do. This would present a considerable problem because of beam divergence. The actual diameter of the beam would have to remain about the same all the way from the earth-based unit to the satellite on which the mirrors were placed. Otherwise, further divergence of the beam would take place after

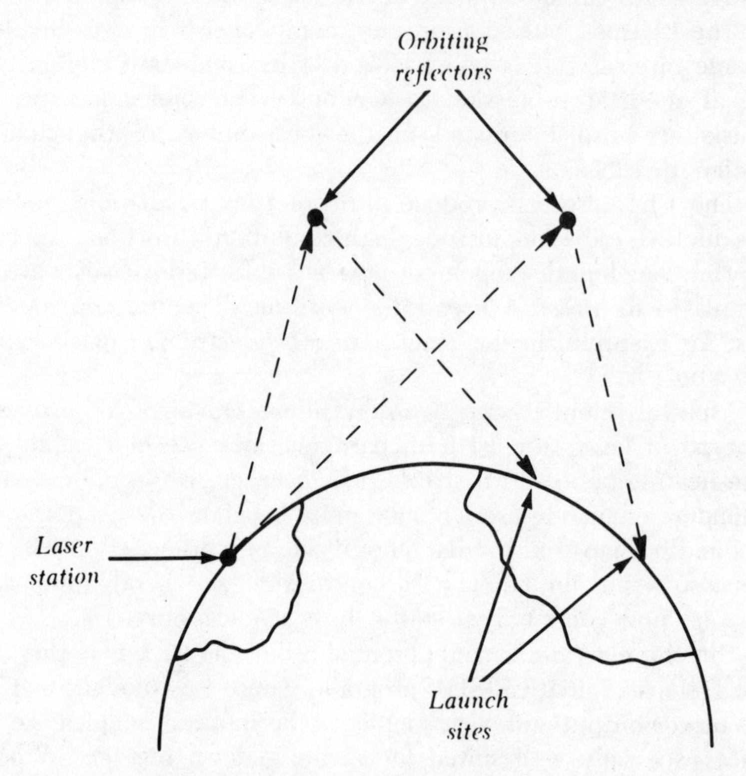

Fig. 7-5. Orbiting reflectors might be used to guide lasers to missile launch sites.

reflection and this would dilute the intensity of the laser beams when they finally struck their targets. For several reflecting mirrors to be used, the beam diameter would have to be fairly large to begin with. The problem of precise aiming of the beam might be mitigated if the satellite were in geostationary orbit, but this would necessitate its being at an altitude of 22,300 miles, and the laser would have to travel about twice this distance before reaching the missiles. Satellites in lower orbits would be moving and so would the missile targets, making the problem of aiming the mirrors and ground-based laser apparatus enormously complex.

THE ALUMINUM LASER

The Reagan administration began serious efforts to deploy a new, high-power, lightweight laser for military use in 1987, to be ready by the early 1990s. This is the aluminum laser, code-named Alpha and scheduled to fly into space as part of the Zenith Star anti-missile laser test. All such programs and systems have code names. The aluminum laser is a chemical device that gets its energy from combustion of fuels similar to the fuels used in rocket engines. This is how the necessary heat is obtained for taking advantage of the emission characteristics of a material that is solid at room temperature.

The initial tests were conducted simply by letting the fuel flow in the device. The testing was done at the Air Force Weapons Laboratory in Albuquerque, New Mexico. This is where the SDI program is managed. The laser is built by TRW, Inc. Tests have been conducted in a secluded valley near San Juan Capistrano, California on a full-scale basis, including even vacuum chambers to closely simulate conditions above the atmosphere.

This particular aluminum laser produces 2 megawatts (2×10^6 watts) of power. That is roughly the equivalent of 1,500 large electric space heaters in operation all at the same time, and confined to a narrow beam of radiant energy. The device has caused some controversy because of its expense and its technical history, which has included several problems, but testing continues at this time. Most of the technical details have of course been kept secret.

LASERS TO GUIDE MISSILES

A laser beam, especially in the near infrared where the energy will penetrate through the atmosphere with minimal attenuation, may be used to "illuminate" a target so that heat or light sensors can be employed for homing. The intensity of such a laser beam need not be anywhere near sufficient to cause damage, and may therefore be several orders of magnitude lower in power than those contemplated for direct destruction.

The principle is quite simple: the laser is aimed at the object designated the target, and the homing device on the "destroyer" missile guides that missile to the target, whereupon the target is destroyed (FIG. 7-6). This technique is being perfected for use in aircraft for air-to-surface missiles, as well as for air-to-air, surface-to-air and surface-to-surface guided missiles.

The problem is, as with any laser, that a line of sight is needed for the system to be effective. Clouds and other obstructions will foil the device. Reflectors at the target will reduce or eliminate the "illumination" caused by the laser. This makes it essential that the lasers be in the invisible wavelengths, so that the illumination cannot be seen with the eye. Devices of this kind, while technologically sophisticated and appealing to some for that reason, are difficult to use properly for the same reason. Someone must operate the equipment and they must have training. Each soldier must have a thorough knowledge of the equipment being used, and as the equipment itself gets more sophisticated, so must the training be to the soldier who operates it. This costs money and takes time. It is clear that, even with training, many soldiers cannot effectively operate these type weapons. In some cases, apparently, old-fashioned manual missile launchers and even heavy artillery have proven just as good or better than laser-guided weapons. Simply being sophisticated does not

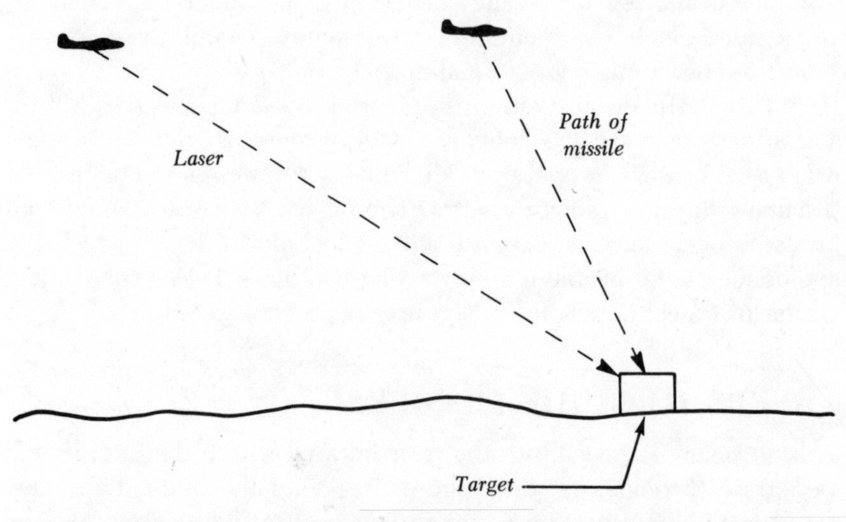

Fig. 7-6. A laser, emitted from one aircraft, provides illumination of a target for a guided missile, in this case fired from another aircraft.

count for much by itself in a real war. What matters then is not how the war is fought, but who wins.

Part of the advantage of using lasers and computers to guide weapons is that it reduces the number of human beings directly involved in perilous combat situations. From that point it is easy to imagine advancing the technology to the point that all the actual fighting is done by computers against computers. Or, more realistically, by robots against robots. Then an attack might be carried out by invading the enemy's "space" (that part of space over their country at altitudes too high for conventional aircraft, out to, say, 1000 kilometers or 600 miles). Any incursion into our "space" would be countered by a force of flying machines and ground-based laser weapons, all unmanned, so that destruction of the machines would not entail any risk to human life. Such a battle, which might look like a scene from the movie *Star Wars*, would be tremendously expensive, but would cause the adversaries to think twice before hurting or killing people. Once one of the countries' forces was defeated, and the battle decided, a negotiated settlement might be hammered out when one or the other country's leaders finally exclaimed, perhaps at the same moment, "Uh-oh, now we might actually have to send in *soldiers*!"

If that is how a war is to be conducted it would be much cheaper to have the two leaders sit down in a video game with the vanquished having to down a shot of vodka or whisky after each individual battle. The loser of the war would just end up under the table.

DESTRUCTIVE POWER OF LASERS

High-powered lasers can be used to cause outright destruction, in addition to just shooting down missiles. A good example is the scenario of a laser being used to destroy an oil refinery. Once started, the fire would likely perpetuate itself. The heat from the laser might be used to ignite some fuel in a part of the refinery where a fire or explosion would be especially likely to spread. You can imagine for yourself what would happen after that.

Lasers might similarly be used to start fires in dry, grassy or wooded areas. The smoke from such fires would reduce visibility and make further use of lasers less effective, but any variable can be used either to advantage or suffered as a disadvantage. Smoke might make excellent cover for a ground invasion, for example.

Laser weapons could be used to cut power lines, destroy roads and railroad tracks, and blow up aircraft while they are still on the ground. All of these applications would require lasers of extreme power.

LOW-POWER USES OF LASERS IN WARFARE

Lasers having less power, not capable of starting fires or explosions or of melting asphalt roads, are still useful in warfare. We have already seen one example, in the guiding of missiles to their targets. However, even lower power devices might be used, for example, to disable or distract visual tracking gear.

There have been reports of pilots being temporarily blinded by what could have been a well-aimed, low-powered laser beam. Just a few milliwatts of laser power are necessary to create a beam that looks as bright as the sun, even from a great distance away. Such a brilliant beam would be distracting, taking a person's attention even if it were not directly observed. This could cause human error in such operations as flying an airplane. If the beam were observed directly it might cause temporary blindness of the sort that happens when you go outdoors on a sunny winter day and then come back inside. Higher-powered lasers are able to cause permanent eye damage.

Reconnaissance cameras can also be "blinded" by a bright light. This is similar to the effect of the sun "washing out" a camera exposure or a television picture. Such a light can render the visual equipment less sensitive or even useless. A good example is the use of a ground-based laser or lasers to blind a satellite camera, making it impossible to resolve pictures. Large lasers might be set up in or near missile bases and other strategic places, as well as in "dummy" locations where no weapons are kept, the "dummy" installations serving to confuse the enemy.

An especially interesting application of low-power and medium-power lasers in the military is for communications purposes. A laser beam could be aimed at clouds, mountains, or simply into the air, and the scattered light picked up and demodulated to retrieve signals. The equipment, if properly designed, might not look like communications apparatus at all, and the enemy might think it was for some other purpose. Such equipment would have limited range but would be very easy to set up and move around, making it ideal for use by ground-based armies. The laser could serve the dual purpose of being modulated for communications and bright enough to blind aircraft pilots or cameras. They could even serve to guide missiles if they were in the infrared range.

Low-powered lasers are commonly used as rangefinders, to determine the distance to a target. This is accomplished by sending a thin beam, either in the visible or infrared wavelength range, to a target, where it is reflected and the returning beam detected by means of a photocell. The device is pulsed and connected to a timer that auotmatically displays the range by measuring the time required for the beam to complete the circuit. This time interval may be measured by a clock oscillator, or by counting the number of pulses that occur between the laser emission and the returned beam.

A similar device is employed to simulate being hit, or "shot in training exercises." Soldiers wear devices on their uniforms that detect laser emissions of a certain wavelength. If such emissions are detected, the soldier is considered eliminated. His weapon, armed with a laser that activates other soldiers' target designators, is disabled by means of a central computer. A block diagram of such a system, with either side being designated by lasers of different frequency, is shown at FIG. 7-7. All the soldiers are designated by a certain code that is transmitted to the central computer if they are hit. Thus the computer knows which soldiers are "casualties" and which are not, and which weapons are to be disabled. Such a system

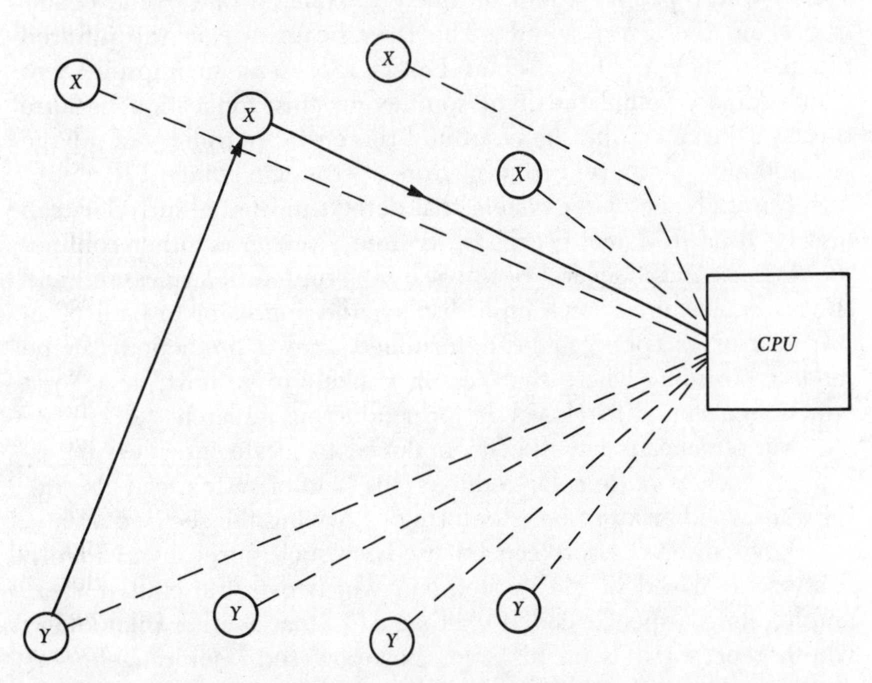

Fig. 7-7. Soldiers (X and Y) are equipped with laser guns and sensors on different frequencies. Here, one of the Y army has hit one of the X soldiers. A signal is sent to the Central Processing Unit (CPU) to indicate the hit.

is not completely realistic because it does not take into account the possibility of a soldier getting hit and not dying. Also, weapons may malfunction in the hands of a perfectly active soldier. But basically the system gives a good indication of whose tactics are working better under the particular conditions of battle. Thus strategies may be put to the test in the field without involving any risk, either of people or of territory, and the cost is fairly low.

Laser rifles using gallium-arsenide devices are commonly used for developing marksmanship. The laser is simply mounted on the barrel of the rifle. The pulses from the laser simulate bullets being fired, except there is no kick. The rifle makes a popping sound to simulate being fired. Laser guns may also be used to simulate real guns on ships, in aircraft, in tanks and other motorized vehicles.

LASERS FOR SURVEILLANCE

Laser beams can be used to determine the presence of trespassers. Some burglar alarm systems operate using lasers and detectors, placed in locations that are unknown to an intruder (and possibly even to the owner of the property). The beams criss-cross the open spaces where people would be likely to walk. If one of the beams is broken, the alarm sounds. The laser beams are in the infrared range, so they are not visible. Lasers represent an improvement over ordinary collimated light sources for this application because the laser beam can be shone around the entire periphery of a large secured area, using reflecting mirrors at strategic points (FIG. 7-8).

Naturally, any laser system that detects unwanted intruders can also be used to detect escapees, as from a prison or other confinement area. And if several beams are interrupted in succession, and their locations are shown on a display, the approximate path of an intruder or escapee can be determined, and a prediction can be made concerning where that person is likely to go next, or at least which area should be closed off for conducting a search.

Various means have been employed to locate intruders or escapees, such as pulse color coding. This kind of system can also determine whether a person is entering or leaving the secured area.

Low-powered lasers can be used for such purposes as identification (a "friend or foe" system) in which every aircraft, for example, emits a specific coded laser signal so that another pilot knows whether or not it is on his side. Low-powered laser rangefinders make good altimeters in clear weather, and can be adapted to indicate ground speed for aircraft and relative velocity for ships, aircraft, and even spacecraft.

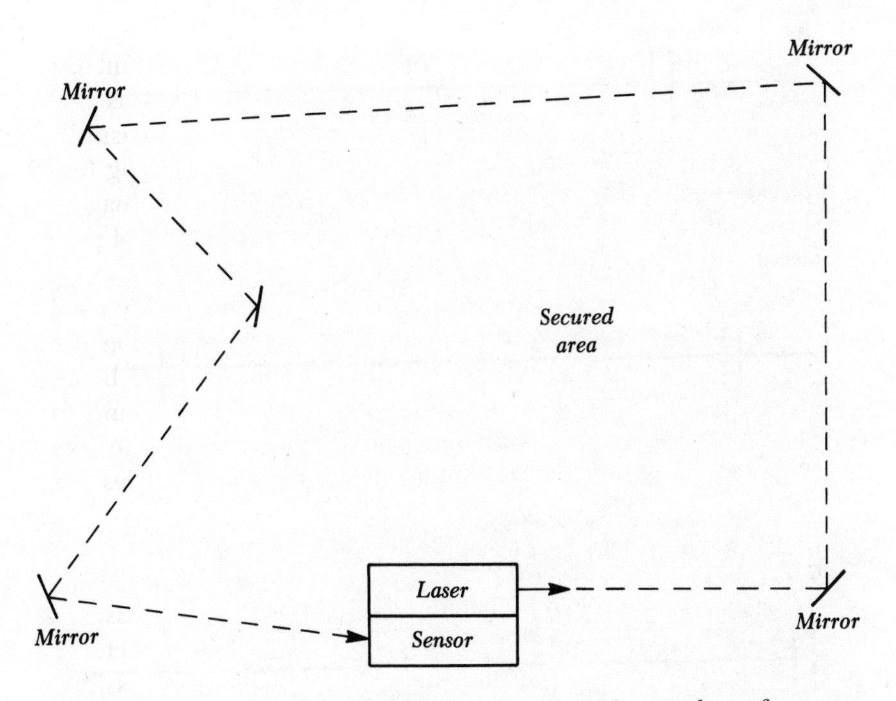

Fig. 7-8. Mirrors and a laser sensor are used to secure the periphery of an area.

Low-powered lasers are, of course, best suited for use above the atmosphere, as with high-power lasers, simply because bad weather cannot foil the operation of the device. In the vacuum of space, a low-powered or medium-powered laser, collimated so that the output consists of essentially parallel rays, will travel vast distances with little attenuation. Such a system might even be used for communicating with spacecraft on the Moon or other planets. Moreover, any wavelength can be used; there are no stop bands in which the atmosphere blocks the energy. Since light has such a short wavelength and because each type of laser has readily identifiable emission bands or frequencies, lasers can be used to measure high speeds by virtue of the Doppler effect. An approaching source or target would generate a "blue shift" and a receding source or target, a "red shift." The displacement (FIG. 7-9) would indicate the radial speed. Other components of the speed could then be determined by parallax observation.

LASERS UNDERWATER

Water is relatively transparent to certain wavelengths of light, although it constitutes a barrier for radio signals, except in the very low frequency range. This fact makes lasers a practical way to com-

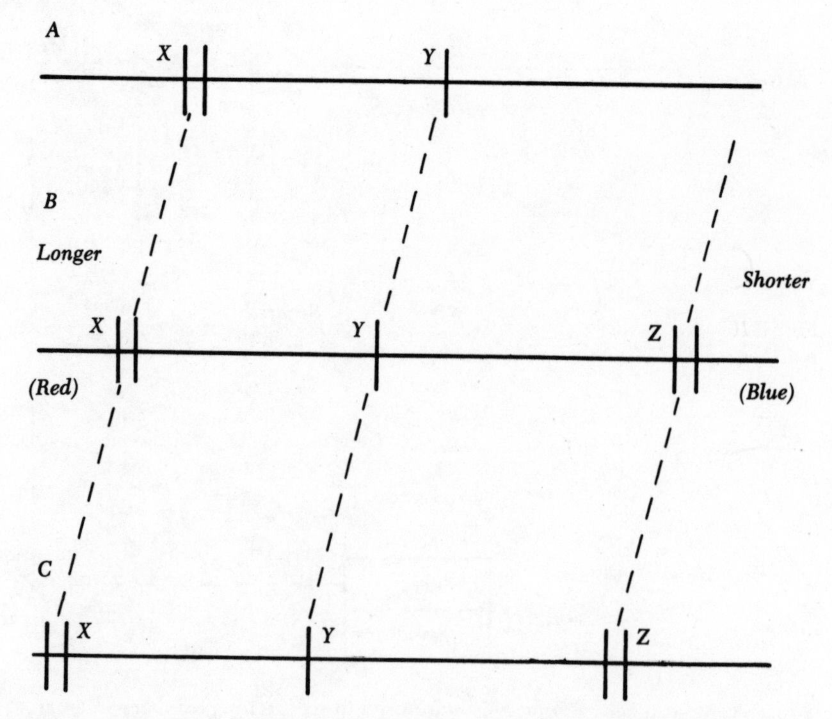

Fig. 7-9. (A) blue shift; (B) red shift. The emission bands are indicated by X, Y and Z (X and Z are double lines). (A) the pair of lines Z has been shifted out of the visible range.

municate between submarines, as well as for target location and guidance systems. The extremely low frequency (ELF) radio signals, with transmitters on land, designed to resonate with the entire Earth, can send signals only at very low rates because of the narrow bandwidths of their modulators. With visible light there is no such limitation, as we have seen.

In relatively clear water, a laser beam might travel a long distance before dying out. In clouded water the attenuation problem would be more serious, although periscopes could be used to transmit the laser beams above the surface (FIG. 7-10). Such beams would be very difficult to even detect, let alone intercept.

Underwater, the principle of total internal reflection (FIG. 7-11) might be employed to provide communication around obstructions such as hills on the bottom of the sea. The detector would be aimed at the image point on the surface. A sensitive receiver would be necessary because the light from such a reflected source would appear to scintillate and messages might become quite garbled because of this "fading."

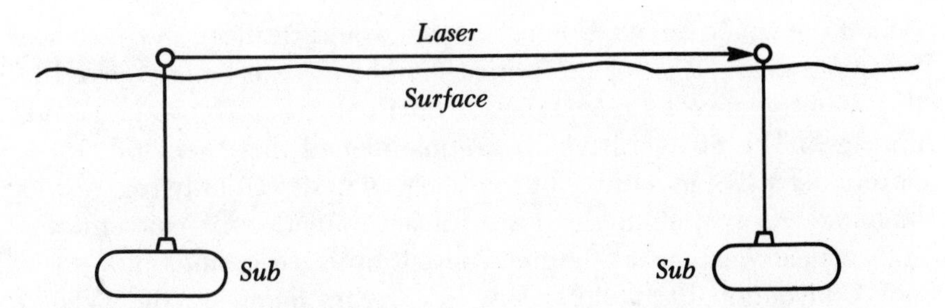

Fig. 7-10. A laser may be used with periscopes for communications between submarines.

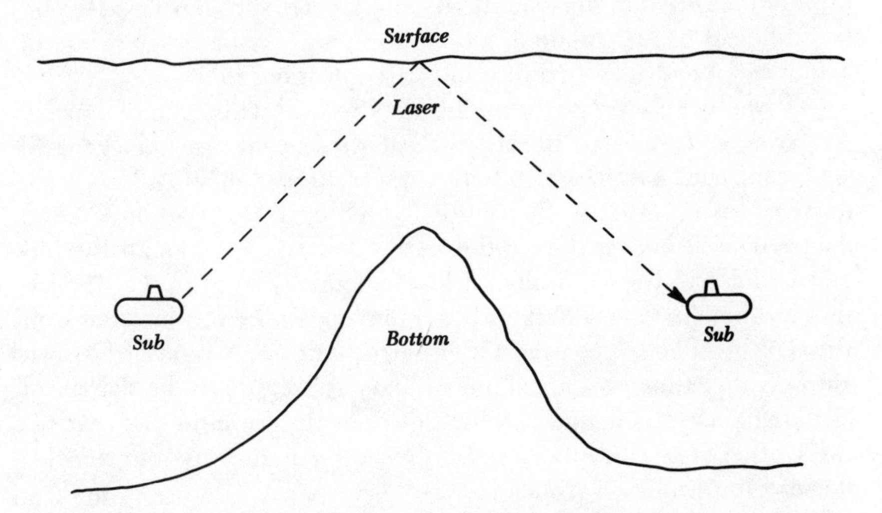

Fig. 7-11. Laser beams may be reflected from the surface for communications underwater between points not on a line of sight.

It is possible, although doubtful, that lasers could be used at high power levels as destructive weapons underwater at close range.

RATIONALE FOR SDI

Why do we really want to have a Strategic Defense Initiative? We do not often say what we really feel about such things directly. It might seem too harsh, but it is a serious matter.

The *status quo* holds that the government of the Soviet Union, in spite of propaganda to the contrary, has serious imperialistic designs. They are furthermore in control of the world's largest land mass and maintain the world's largest military machine. Their ultimate goal is believed to be to control the remainder of what they do not already occupy militarily. Most people who have much perception of life in the Soviet Union feel this would result in more-or-less complete elimination of the personal freedoms we have enjoyed in this country since its inception and in England since the late Middle Ages. The Soviets have always maintained that their military is purely defensive in nature, but military analysts can only regard this huge and overwhelmingly powerful accumulation of conventional and nuclear weapons as the main threat to peace in our time.

Conventional strategists since the years following the Second World War have warned that a relaxation of our nuclear offensive capabilities would result in a "nuclear blackmail" situation, in which merely the threat of nuclear attack by a greatly superior force would be sufficient to intimidate any opposition and enable the Soviets to achieve any end they had in mind without opposition.

It was concluded that the only way to deal with such doomsday scenarios was to make it evident that there could be no victor in such confrontations in which both sides possessed more or less equal nuclear power. This is the Mutually Assured Destruction (MAD) doctrine, well known during the sixties and later. Although the Soviet Union had been our ally, at least nominally, during the war (although they had been Hitler's ally immediately before his attack on Russia) the effective conquest and permanent occupation of Eastern Europe after the war alerted our military strategists to the danger of allowing such a vast power which did not disarm after the war, as the United States did, to continue their campaign across the remainder of Western ern Europe.

Thus came about the curious but effective "standoff" known as the "Cold War," which in spite of occasional conflicts, has restrained the major nations from all-out war since 1945—a remarkably long period of relative peace in terms of world history. We have in effect blackmailed ourselves into peace. This turned out to be a boon for those fortunate enough to be on the American side of Europe at the end of the war, and unfortunately, conditions little removed from slavery for those of east Europe and the Soviet Union. Ask any survivor of the Bataan Death March (there were not manv) if he thinks slavery is a joke.

A nuclear holocaust is therefore as unlikely today as it was at the evolution of this strategy, and no doubt will continue to be unlikely as long as we adhere to this status quo. But that reasoning has never satisfied everyone, and there is some possibility, although the precautions are extreme, that either through error or design, someone could some day get nervous or go insane and simply "pull the trigger," thus starting something irreversible that no one wanted.

The SDI program, or "Star Wars" as it has come to be inevitably called, is further insurance against such an eventuality, rather than just another new weapons system as it is often perceived by the uninformed. It is a system which would be capable of destroying offensive nuclear weapons, usually imagined as a powerful laser "space ray gun," which would incinerate missiles in flight before they could reach a target. The advantages to having such an option to prevent the unthinkable should be apparent. It is therefore revealing that the major opposition to its development here comes from the Soviet Union, although all available evidence suggests that they are proceeding full speed with their own Star Wars program. I am not sure how safe I would feel knowing the Soviet Union had such a system in place, while we in the United States had allowed ourselves to bargain the program away.

At any rate, no nation has yet developed such a vastly complex system. The difficulties involved in developing and deploying SDI are enormous, and much information surrounding this program is no doubt "classified" by our government. However, as in most involved defense research projects, there will also be what are called *spin-offs*—civilian uses found for the technology developed along the way. Some laser designs intended for SDI are already being used in medicine.

Lasers aren't the only weapon of interest to SDI, of course. Missiles may just as well be blown up by means of some other type of bomb, or detonated early, while still at great altitudes. This would not require the use of high-powered "ray guns" at all.

Finally, the simplest analogy which best illustrates the effect of a successful SDI program would probably be the familiar old western scenario of two gunfighters facing each other in a showdown on main street at high noon. Except that now one of the two has equipped himself with a bullet proof vest.

Bibliography

Bertolotti, Mario. *Masers and Lasers: An Historical Approach* (Adam Hilger, Ltd., 1983)

Brown, Ronald. *Lasers: Tools of Modern Technology* (Doubleday & Company, 1968)

Carroll, John M. *The Story of the Laser* (E. P. Dutton & Co., 1970)

Hallmark, Clayton C. and Horn, Delton T. *Lasers: The Light Fantastic* (TAB Books, Inc., 1979)

Heckman, Philip. *The Magic of Holography* (Atheneum, 1986)

Muncheryan, H. M. *Principles and Practice of Laser Technology* (TAB Books, Inc., 1983)

Stehling, Kurt R. *Lasers and their Applications* (World Publishing Co., 1966)

Index

T

U

V

W

X